Vedic Tricks Made Mathematics Easy

Vedic Tricks Made Mathematics Easy
The Ultimate Guide for Swift Mental Calculation to Perform Better in Exams

Self-Published
(MAHAVIR SINGH)

Second Edition

© Copyright, 2024

www.sparktriangle.com

Vedic Tricks Made Mathematics Easy

The Ultimate Guide for Swift Mental Calculation to Perform Better in Exams

MAHAVIR SINGH

Insights and Inspirations on Vedic Mathematics

"Vedic Mathematics is a gift to the world that simplifies complex calculations, making learning an enjoyable and enriching experience."

- Swami Bharati Krishna Tirthaji

"Vedic Mathematics is not just a collection of tricks; it is a philosophy of simplifying and understanding the nature of numbers."

- Nicholas Khatchadourian

"The ancient system of Vedic Mathematics offers profound insights into the simplicity and efficiency of mathematical calculations."

- Shakuntala Devi

"The methods of Vedic Mathematics provide a refreshing alternative to conventional arithmetic, unveiling a world of mental agility."

- James T. Glover

"Exploring Vedic Mathematics opens up new horizons in mathematical thinking and problem-solving."

- S.G. Dani

"Vedic Mathematics is a testament to the intellectual achievements of ancient India and its relevance in today's world."

- Dr. M.D. Srinivas

CONTENTS

Chapter – 1...1

Introduction

Chapter – 2...8

Secretes of Swift Calculation at Fingertips

- Perform Multiplication from 6 to 25 with Fingers Tricks
- Quick Counting From 1 to 99 with Fingers
- Quick Addition & Subtraction with Fingers
- Abacus Trick for Quick Calculations

Chapter – 3..27

Vedic Tricks of Number Addition

- Find the Digit Sum (Using 9-Point Circle)
- Number Addition Completion or Non-completion (Using 10 Point Circle)
- Additions by Making Multiple of 10, (Using Sutra 'By the Deficiency')
- Numbers Addition (Start Left to Right)
- The Addition of Numbers, Where the Unit Digits of the Numbers Add Up to 10.
- Addition of List Number (Single Digit Pure Number)

- o G) Divide Any Numbers By 50 (Using Doubling and Divide by 100 Method)
  - o H) Divide Any Numbers By 25 (Doubling Twice Then Divide by 100)
- *Multiplication of Numbers with Patterns that Add to 10.*
  - o A) Squaring a Number, Ending with 5.
  - o B) Multiplication of Numbers, Where "First Digits are Same, and Last Digits Add to 10"
  - o C) Multiplication of Numbers, where "First Digits Add to 10" and Last Digits are Same".
  - o D) Multiplication of Numbers, Where the First Digits of the Two Numbers are (n) and ($n+1$) and Last Digits Add to 10.
  - o E) Multiplication of Numbers, Where Last Digits of Both the Numbers is 5 and the Difference of Two Numbers are Equal to 10.
- *Multiplication of Any 2-Digit Numbers*
  - o Using Vedic Sutra "Urdhva-triyagbhyam", which means "Cross-wise and Vertical".
- *Multiplication of Numbers in Series of "1"*
  - o A) Multiplication of Two Numbers Ending with Last Digits "1"
  - o B) Multiplication method of two-digit numbers by 11.
  - o C) Multiplication of 3-digit numbers by 11
  - o D) Multiply numbers with larger digits by 11.
  - o E) Multiplication of 2-digit numbers by triple digit "111-series".
  - o F) Multiplication of 3-digit numbers by triple digit "111-series".
  - o G) Multiplication of 4-digit numbers by triple digit "111-series".
  - o H) Multiplication of 5-digit numbers by triple digit "111-series".

Chapter – 6 111

Quick Technique of Squaring Numbers

Chapter – 7 122

Vedic Method of Division

Cube-Roots of Perfect Cube

- Cube-roots of perfect numbers using Vedic sutra "Vilokanam" which means 'By mere Observation'
- Observation table
- Quick tricks of finding out the cube root of 5-digit numbers

Vedic Square and Number Patterns

- Unique Patterns Generated from Vedic Square

Magic Squares

- Steps for Creating a Magic Square of n x n
- Steps for Creating a Complementary Magic Square

Special Section for Competitive Exams

- Number Theory
- Odd, Even, Square, Cube, and Prime Numbers
- Euclid's division lemma
- Divisibility of Numbers
- Digit Sums
- Amazing Short Tricks of Simplifying Various Mathematical Puzzles, in Competitive Exams
- Solving Complex Mathematical Equations Quickly
- Addition Tricks for 10 - Continuous Natural Numbers.

- Amazing Tricks of Simplifying Surds and Indices
- Division of Infinite Series of Surds
- Basic Multiplication, Division, Power, and Exponent rules of indices

Sutra in Vedic Mathematics with Applications

- 16-Vedic Sutras (Aphorisms) & Applications
- 13-Sub-sutras (Corollaries) & Applications

Swift Mental Calculation Tricks

- Mental Tricks of Addition
- Mental Tricks of Subtraction
- Mental Tricks of Multiplication
- Mental Tricks of Division
- Mental Tricks of Fractions

- Exercise For Practice

- Author Connect
- Feedback and Suggestions

Preface

In creating this edition, I express profound gratitude for the invaluable feedback from readers of the first edition. Your input has played a pivotal role in enhancing the overall quality and relevance of this edition, and I extend my sincere appreciation for your thoughtful contributions.

Your insights have shaped this resource, guiding the addition of new chapters, examples, and improved graphics. This collaborative effort aims to be a refined, reader-centric source addressing diverse learner needs.

This special edition focused on simplifying vedic tricks for complex mathematical problems, this edition integrates numerous examples and graphics. The goal is to enhance the reader's learning experience, providing new chapters tailored for competitive exams and various mathematical tricks for quick calculation, be it mental or finger trick calculations.

Recognizing the significance of competitive exams, dedicated chapters have been thoughtfully integrated, offering specialized guidance to those preparing for the exams ahead. I've tailored the content to provide targeted assistance for those navigating competitive assessments, ensuring a clear and accessible comprehension for all readers of age 8 years and above.

May these enhancements enrich your learning experience, offering valuable insights into mathematics intricacies, serving as a reliable guide for academic pursuits and competitive aspirations.

Dedication

To the resilient individuals embarking on the challenging path of exam preparation, to the ardent lovers of mathematics, and to those constrained by financial barriers in expanding their educational resources - this book is dedicated to you, aiming to be a steadfast companion on your journey.

As we forge ahead, let us draw inspiration not only from the rich mathematical content within these pages but also from the collective spirit of perseverance that binds us. Each chapter of the book unveils the artistry of 'Vedic Tricks for Mental Maths'

May this edition serve as a beacon of knowledge and empowerment, resonating with the enduring wisdom of the ancient insights. Here's to embracing the elegance of vedic mathematics and the pursuit of enlightenment, unbounded by limitless possibilities.

Acknowledgments

I would like to express my sincerest gratitude to the readers of the first edition, whose valuable feedback and suggestions played a pivotal role in shaping the much advance version of second edition. Your unwavering support and insightful comments have been instrumental in making this edition a game changer for new readers. Your feedback has allowed me to delve deeper into the subject matter, refine concepts, and provide even more clarity and practicality in the vedic tricks of calculation. Through your contributions, this edition shines with a new level of excellence and effectiveness. Your willingness to share your thoughts and experiences has not only benefited this book but also the wider community of readers seeking to enhance their mathematical skills.

I am humbled and grateful to have your continued support and encouragement in this journey. Your trust in my work has inspired me to deliver an edition that goes above and beyond, offering immense value and empowering individuals to unlock their full potential. Once again, thanks to all the readers of the first edition for your invaluable contributions. Your suggestions have truly made a difference, and I am immensely grateful for your role in making this edition a game changer for new readers. I hope the revised content brings you even more joy, enlightenment, and success in your mathematical endeavours.

Chapter – 1

Introduction

Welcome to the enhanced and enriched second edition of my earlier book *Miracles of Vedic Tricks for Mental Maths*. In this iteration, the book undergoes a transformative evolution, meticulously crafted to elevate your learning experience. Embracing the invaluable feedback from our readers, we've incorporated an array of enhancements. More examples illuminate the path to understanding, while graphic representations of formulae bring clarity to complex concepts.

New methods and tricks, focused on competitive exam needs, serve as your strategic toolkit. With each turn of the page, you'll benefit from a comprehensive range of knowledge designed to empower your journey of vedic mathematics. Join us to this upgraded version of the book, where the synergy of expertise and innovation converges to redefine your learning adventure.

The journey of Vedic to modern calculation spans eras, showcasing a rich evolution in mathematical thought. Rooted in ancient India, the Vedic period 1500 BCE to 500 BCE witnessed the development of the Sulba Sutras (considered to be appendices to the Vedas), mathematical treatises that laid the foundation for geometric and algebraic concepts. Vedic scholars displayed remarkable insights into arithmetic, geometry, and algebra, fostering a deep understanding of numbers. As centuries unfolded, mathematical knowledge travelled across cultures and regions, contributing to the global development of mathematics. The advent of numeral systems, including the influential Hindu-Arabic numerals, marked a significant shift towards more efficient and versatile calculations.

Innovations by mathematicians like Brahmagupta and Aryabhata further propelled mathematical knowledge. With the Renaissance and subsequent eras, the foundations laid during the Vedic period converged with advancements in Europe, leading to the modern mathematical framework we employ today, influencing fields beyond mathematics, such as physics, engineering, and computer science.

The journey from Vedic to modern calculation is a testament to the enduring impact of ancient wisdom on the continuous refinement of mathematical understanding.

The first method of counting has been argued to be counting on fingers, stones, or any various kinds of natural material. The abacus is, one of the oldest known counting devices, has a rich historical background. Its origins can be traced back thousands of years to ancient civilizations such as Mesopotamia, Egypt, and China. The abacus served as a tool for performing basic arithmetic calculations before the advent of modern numeral systems.

Vedic maths is an ancient system of mathematics basically originated in India. Vedic maths offers a unique and holistic approach to mathematics, provides fastest mental calculations, quick problem-solving techniques, and mathematical shortcuts. It's absolutely based on rational and logical reasoning.

I, once again welcome you to the realm of extraordinary mathematical prowess. In this inventive book, we delve into the ancient secrets of Vedic mathematics and unleashing the unpredictable power of lightning-fast mental calculations. Prepare to embark on a journey that will change your approach to numbers, unveiling a realm of limitless possibilities.

Throughout the history, humankind has strived to simplify and expedite the art of mathematics. The miracles of Vedic tricks for mental maths are manifold. Imagine calculation of multiplications, divisions, square roots, and cube roots without relying on pen and paper or electronic devices.

The vedic tricks can swiftly help you solving equations, unravelling complicated patterns, and deciphering difficult mathematical puzzles, all with the astounding speed and accuracy. The benefits of mastering mental maths, go far beyond mere arithmetic proficiency. Cultivating these skills enhances logical reasoning, fosters creativity, and cultivates a deep sense of numerical intuition.

In Vedic mathematics, you will witness the transformative power it holds, not only in mathematical calculations, but also in everyday life situations that call for quick solutions. So, prepare to embark on this extraordinary journey of discovery of new mathematical tricks and embrace the power of lightning-fast calculation skills.

The wonders that lie ahead will forever change the way you perceive the world of numbers and arithmetic calculations. Let us begin this captivating expedition, where ancient wisdom meets modern brilliance, and the art of mental maths becomes a key to unlocking a universe of mathematical miracles.

The term vedic is originated from the Sanskrit word "Veda, which means 'knowledge", the vedas is derived from the root "Vid", meaning to know without limit. Vedas are a large body of religious texts originated in ancient India. Composed in vedic Sanskrit, the texts constitute the oldest layer of Sanskrit literature and the oldest scriptures of Hinduism. There are four vedas namely Rigveda, Yajurveda, Samaveda and the Atharvaveda.

The principles and techniques of vedic mathematics are based on Vedic sutra, which were primarily taken from the parisiṣṭa (a supplementary text or appendix of the Atharvaveda). The Atharvaveda is a collection of mantras and formulas, which deals with the branches like engineering, mathematics, sculpture, medicine, and all other sciences. The word Veda covers all sakhas (branches) known to humanity. It is a repository of all knowledge, fathomless, ever revealing as it is delved deeper.

The great mathematician the *Shankaracharya Jagadguru Bharati krishna Tirthaji Maharaj*, had rediscovered and made significant contributions to vedic mathematics. He is a scholar and spiritual leader, and has widely regarded for his efforts to compile and systematize the techniques of vedic mathematics. In the early twentieth century, he wrote the book "Vedic Mathematics" where he presented various mathematical principles and methods. His work played a significant role in reviving interest in vedic mathematics and bringing it to a wider audience. His efforts in organizing and popularizing these techniques have had a profound impact on the understanding and spread of vedic mathematics.

The vedic maths enhances ability to speed calculations of numbers with high accuracy. It promotes the mental calculation, increases creativity and help in forming patterns and creating their own formulas, because it provides multiple ways to solve a problem, reduces the burden of memorizing difficult formula, which cultivates the interest in mathematics.

Indian Prime Minister, *Shri Narendra Modi,* during his *88th* episode of *'Mann Ki Baat'*, has urged parents to teach Vedic mathematics to their children. He said: *"I, would like all parents to teach Vedic math to their children. With this, their confidence will not only increase; the analytical power of their brain will also increase and yes, whatever little fear some children have about mathematics, that phobia will also end completely".*

He rightly said that we must teach Vedic mathematics to our children; it would be one of the most precious gifts that all parents can give to their children, making their tomorrow much brighter. As the Vedic tricks accelerate mental abilities and developed a problem-solving mindset by understanding the versatile nature of Vedic mathematics. They can learn how to apply the techniques to a variety of mathematical challenges. Once, they become master in vedic tricks, they can solve most complicated mathematical problems in few seconds only.

You can apply your newfound skills to everyday situations, whether you're a student or professional, mastering of Vedic tricks will empower you with a valuable skill set that will save time, enhance your mathematical abilities, and boost your confidence.

The vedic mathematics is gaining its popularity and recognition globally, due to its ease of calculation and capability of solving complex problem with accuracy in a lesser time. Embrace the wisdom of the ancient and unleash your mathematical genius of today.

The book has been written deliberately to students, educators, professionals, or those preparing for competitive exams. Learning magical tricks for complicated mathematics can be a transformative experience to everyone, helping them develop a deeper understanding of numbers.

The special chapters included in this edition can be immensely beneficial for individuals preparing for competitive exams as well as for school-going children. The book will provide invaluable support and knowledge, assisting them on their journey towards success.

I wish you all the very best!

Chapter – 2

Secretes of Swift Calculation at Fingertips

Perform Multiplication from 6 to 25 with Fingers Tricks

It is an interesting tricks of multiplication, where you can use your fingers gestures to multiply numbers from 2 to 25.

This trick is very interesting, which allow computation on fingers to perform mathematical calculations. We can call it hand abacus. The trick is based on the traditional abacus concept but is designed to be carried on the fingers, providing a portable and tangible method for counting and arithmetic.

Each finger typically represents a different place value, and by renumbering the fingers, we can efficiently perform calculations.

Multiply Single Digit Number By 9 Using Finger Tricks

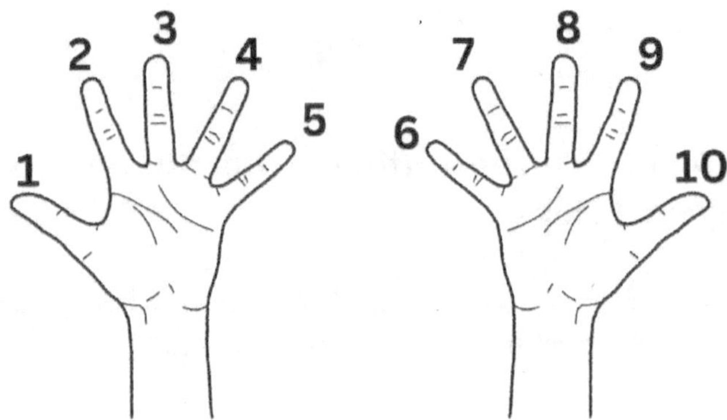

Here are the steps:

- Numbering the fingers from 1 to 10, from left to right.

- Now just fold down the finger with number you want to multiply by 9.

- Then count fingers on the left side (give first part of the result) and right side of the folded finger (give second part of the result).

- Combine the numbers counted on left and right side of the folded finger, that will give the result of multiplication.

Examples

1) Multiply 4 × 9 with your fingers

Simple Steps

- Just *fold down* the *finger number 4* and count the no. of fingers on left and the right of the folded finger.

- Just note down the no. of fingers counted on the left and the right side. i.e. 3 and 6

- The answer is 36. (i.e. 4 x 9 = 36).

2) Multiply 2 × 9 with your fingers

Simple Steps

- Just *fold down* the *finger number 2* and count the no. of fingers on left and the right of the folded finger.

- Just note down the no. of fingers counted on the left and the right side. i.e. 1 and 9

- The answer is 18. (i.e. 2 x 9 = 18).

3) Multiply 6 × 9 with your fingers

Simple Steps

- Just *fold down* the *finger number 6* and count the no. of fingers on left and the right of the folded finger.

- Just note down the no. of fingers counted on the left and the right side. i.e. 5 and 4

- The answer is 54. (i.e. 6 x 9 = 54).

4) Multiply 8 × 9 with your fingers

Simple Steps

- Just *fold down* the *finger number 8* and count the no. of fingers on left and the right of the folded finger.

- Just note down the no. of fingers counted on the left and the right side. i.e. 7 and 2

- The answer is 72. (i.e. 8 x 9 = 72).

Tricks For Multiplying Numbers on Your Fingers (Between 6 To 10)

Here are the steps:

- Numbering the fingers on both hand from 6 to 10 (starting from little fingers).

- Now, just fold all the fingers on each hand, which you want to multiply to each other.

- Count all the folded fingers on both hand (each folded finger value is 10).

- Multiply unfolded fingers of both hand

- Then add two results

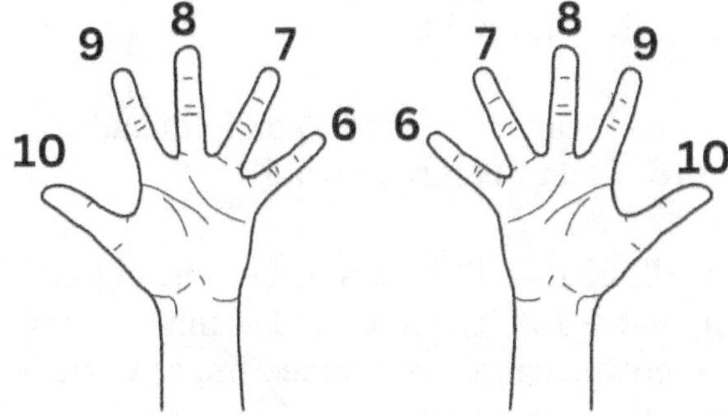

#Rule 1: The value of each folded finger is 10.
#Rule 2: The value of each unfolded finger is 1.
#Rule 3: Multiply unfolded finger on both hand to each other.

#Rule 4: If multiplication of unfolded fingers on two hands results into double digits, then tenth place digit is carry forwarded to left (i.e. one more 10 is added to the result of folded fingers)

Examples

1) Multiply 7 × 8 with fingers

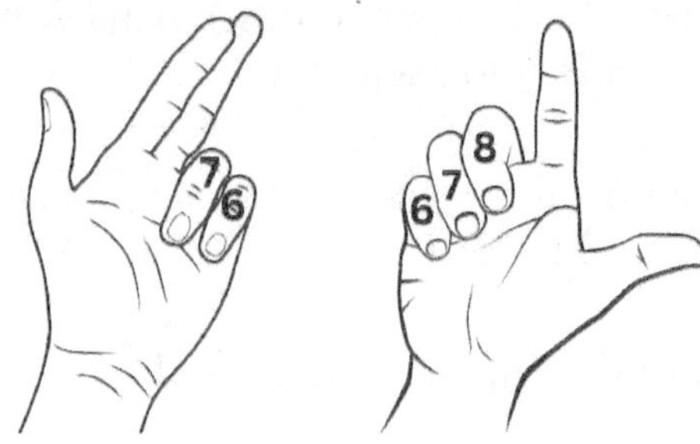

Here are the steps:
- Numbering the fingers on both hand from 6 to 10 (starting from little fingers).

- Now, fold 2-fingers (6, and 7) on left hand and finger (6, 7 and 8) on right hand.

- Count all the folded fingers on both hand (each having value 10). The total folded fingers are (2 + 3) = 5, now because each folded finger value is 10, so the total value is **50**.

- Now, multiply unfolded fingers of both hand 3 × 2 = **6**

- Add the two results, 50 + 6 = **56 Ans.**

Tricks of Multiplying Any Numbers with Your Fingers (Between 11 To 15)

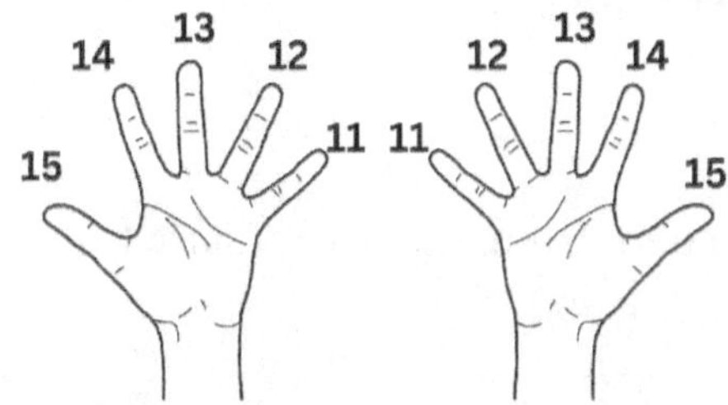

#Rule 1: The value of each folded finger is 10.

#Rule 2: Do not worry about unfolded fingers, they are no more in use.

#Rule 3: Standard value 100, is added to the final result.

Here are the steps:

- Numbering the fingers on both hand from 11 to 15 (starting from little fingers).

- Now, just fold each fingers on both hand, which you want to multiply to each other.

- Count all the folded fingers on both hand (each folded finger value is 10).

- Count & multiply folded fingers on each hand

- Then add two results

Examples

1) Multiply 12 × 13 with fingers

Here are the steps:

Numbering your fingers from 11 to 15, (start from little fingers to thumbs).

Now fold 2- fingers (11 & 12) on left hands, and 3-fingers (11, 12 & 13) on right hand.

Counts the no. of folded fingers (value of each folded finger is 10). Here are total (2+3) = 5 folded fingers, with value 50.

Multiply the folded fingers on both hand
2 x 3 = **6**

Add this multiplication to the total value of folded fingers 50 + 6 = 56.

Finally add standard value 100 to the above result (will give the final answer), i.e. 56 + 100 =**156 Ans.**

Tricks of Multiplying Numbers between 16 to 20 on Fingers.

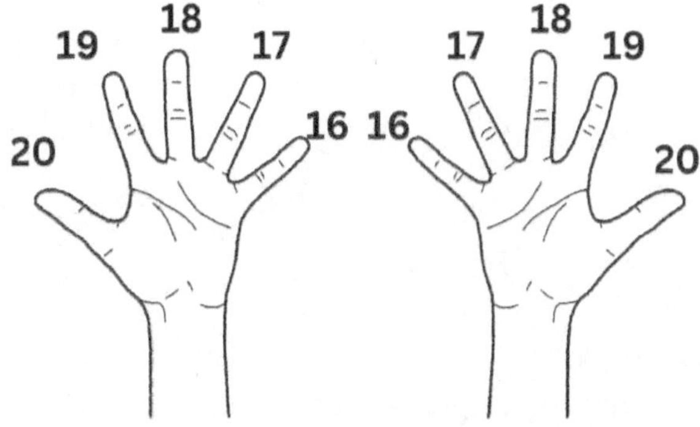

#Rule 1: The value of each folded finger is 20.

#Rule 2: Numbering your fingers from 16 to 20, (begin from little fingers to thumbs).

#Rule 3: Now just fold the finger-numbers of your both hands to which you need multiply two numbers.

#Rule 4: Counts the no. of folded fingers (remember, the value of each folded finger is 20) and note down the total value.

#Rule 5: Count & multiply the folded fingers on both hand and add result to the total value of folded fingers.

#Rule 6: Now add standard value 200 to the above result, that will give the final answer.

Examples: Multiply 17 × 18 with fingers

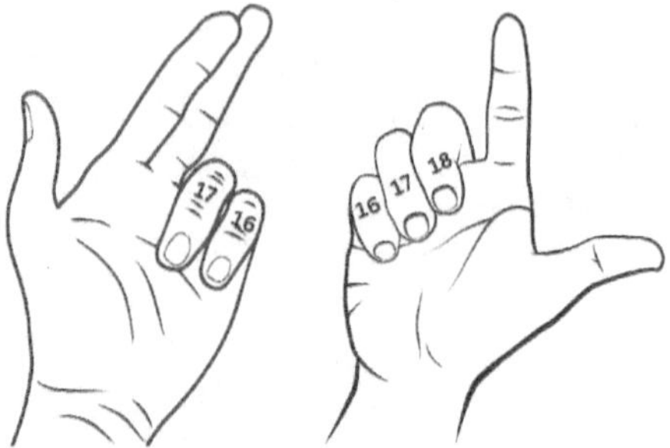

Here are the steps:

- Numbering the fingers on both hand from 16 to 20 (starting from little fingers).

- Now, fold 2-fingers (16, and 17) on left hand and finger (16, 17 and 18) on right hand.

- Count all the folded fingers on both hand (value of each folded finger is 20). The total folded fingers are (2 + 3) = 5, now because each folded finger value is 20, so the total value is **100**.

- Now, count & multiply folded fingers on each hand 3 × 2 = **6**

- Add the two results, 100 + 6 = **106**

- Now add standard value 200 to the above result, that will give the final answer.

- 106 + 200 = **306 Ans.**

Tricks of Multiplying Numbers between 21 to 25 on Fingers.

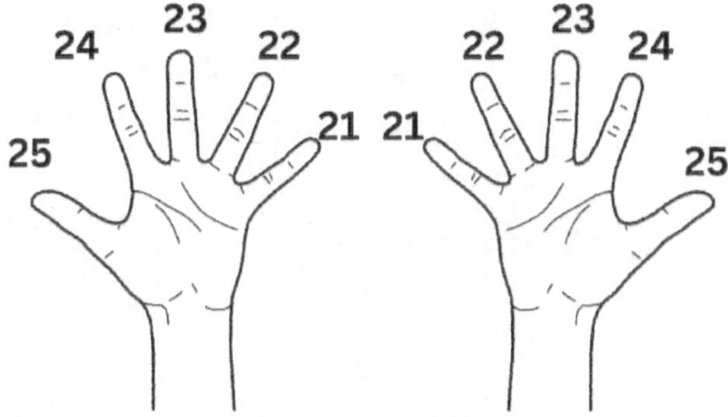

#Rule 1: The value of each folded finger is 20.

#Rule 2: Do not worry about unfolded fingers, they no more in use.

#Rule 3: Standard value 400, is added to the final result.

#Rule 4: Numbering your fingers from 21 to 25, (begin from little fingers to thumbs).

#Rule 5: Now just fold the finger-numbers of your both hands to which you need multiply two numbers.

#Rule 6: Counts the no. of folded fingers (remember, the value of each folded finger is 20) and note down the total value.

#Rule 7: Multiply the folded fingers on both hand and add result to the total value of folded fingers (note: unfolded fingers are not in used for this case).

#Rule 8: Now add standard value 400 to the above result, that will give the final answer.

Examples

1) Multiply 22 × 23 with fingers

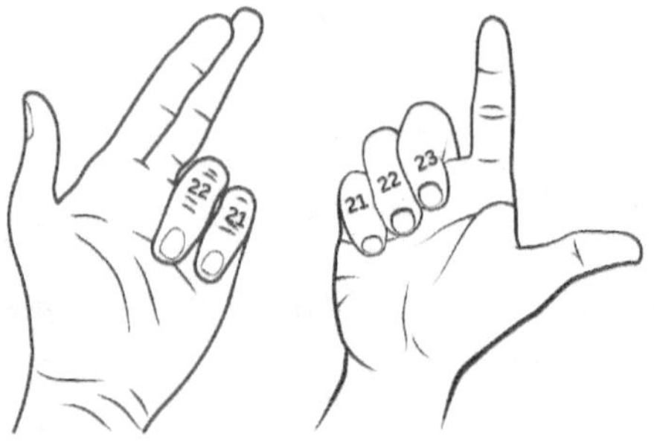

Here are the steps:

- Numbering the fingers on both hand from 21 to 25 (starting from little fingers).

- Now, fold 2-fingers (21, and 22) on left hand and finger (21, 22 and 23) on right hand.

- Count all the folded fingers on both hand (value of each folded finger is 20).

- The total folded fingers are (2 + 3) = 5, now because each folded finger value is 20, so the total value is **100**.

- Now, count & multiply folded fingers on each hand
  $3 \times 2 = \mathbf{6}$

- Add the two results, $100 + 6 = \mathbf{106}$

- Now add standard value 400 to the above result, that will give the final answer.

- $106 + 400 = \mathbf{506}$ **Ans.**

Practice the following multiplications using finger tricks

1) Multiply 21×22

2) Multiply 21×23

3) Multiply 21×24

4) Multiply 22×24

5) Multiply 22×24

6) Multiply 22×25

7) Multiply 23×22

8) Multiply 23×24

9) Multiply 23×25

10) Multiply 24×25

Quick Counting from 1 to 99 with Fingers

#Rule-1: Each fingers on the right-hand represent a number 1 and thumb represent the number 5.

#Rule-2: Each fingers on the left-hand represent a number 10 and thumb represent the number 50.

Left hand fingers: The value of each finger on left hand are equal to 10 and the value of thumb is equal-50. So, the total counts on left hand is 90

Right hand counts: The value of each finger on right hand are equal to 1 and the value of thumb is equal-5. So, the total counts on right hand is 9.

Thus, the total counts of fingers on both the hands are **99**.

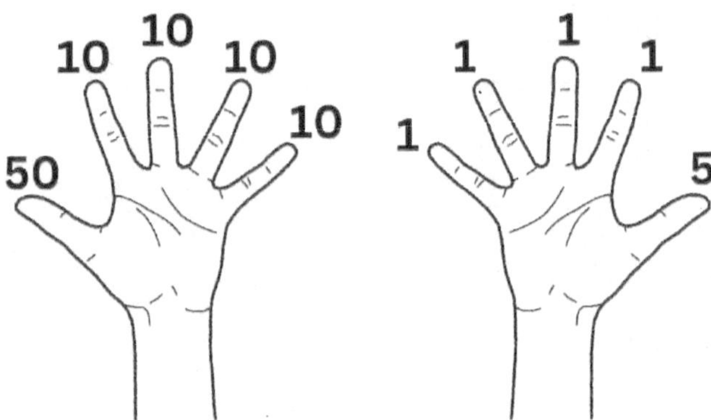

Here's the steps to learn basic number counting on fingers.

Quick Addition & Subtraction with Fingers

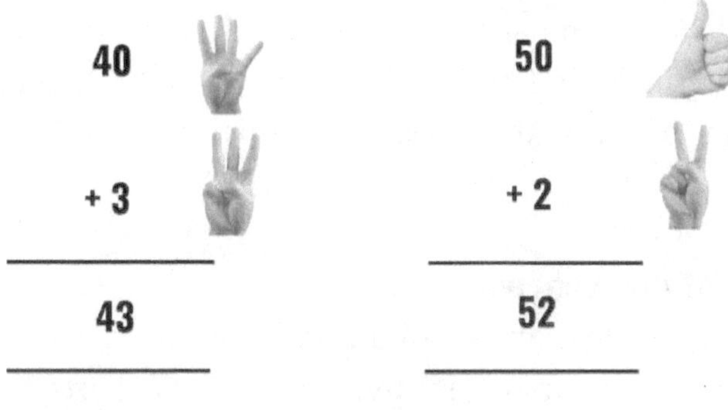

```
 40
+ 3
----
 43
----
```

```
 50
+ 2
----
 52
----
```

```
 50
+ 4
----
 54
----
```

```
 60
+ 4
----
 64
----
```

Subtraction

In this case you just fold the fingers you want subtract, and count the remaining fingers, will give the answer.

Abacus Trick For Quick Calculations

The abacus is an ancient tool used for arithmetic calculations, including addition and subtraction. A simple explanation of how the abacus can be used for these operations:

Structure of the Abacus

An abacus typically consists of a frame with rods, each rod having beads. The frame is divided into two parts by a horizontal bar:

Upper part (Heaven): Each rod has one bead worth 5 units.

Lower part (Earth): Each rod has four beads, each worth 1 unit.

Basic Steps for Addition

Place the first number on the abacus. For example, to represent 456, you would place 4 beads on the hundreds rod, 5 beads on the tens rod, and 6 beads on the units rod.

Now, add numbers digit by digit, start from the rightmost digit (units place). If the result on a rod exceeds 9, carry over to the next rod. For example, adding 7 + 6 in the units place gives 13. You place 3 beads on the units rod and add 1 bead to the tens rod. Continue the Addition: Move to the next digit (tens place), and then to the hundreds place, applying the same carry-over rule.

Basic Steps for Subtraction

Place the larger number on the abacus. For example, to subtract 234 from 456, you first place 456 on the abacus.

Now, subtract numbers digit by digit, start from the rightmost digit (units place). If the bead count on a rod goes below 0, borrow from the next higher rod.

For example, subtracting 6 from 4 in the units place requires you to borrow 1 from the tens place, making the units place 14, and then subtract 6 from 14, leaving 8 in the units place. Continue the Subtraction: Move to the next digit (tens place), and then to the hundreds place, applying the same borrowing rule.

Example-1: Add 123 and 456 on abacus

Step-1: Place 123 on the abacus.
Step-2: Add 456 digit by digit:

Units: 3 + 6 = 9 (no carry needed).
Tens: 2 + 5 = 7 (no carry needed).
Hundreds: 1 + 4 = 5 (no carry needed).
The abacus should show 579.

Example-2: Subtract 123 from 456 on abacus

Step-1: Place 456 on the abacus.
Step-2: Subtract 123 digit by digit start from left to right:
Hundreds: 4 - 1 = 3.
Tens: 5 - 2 = 3.
Units: 6 - 3 = 3.

The abacus should show 333.

Example-3: Subtract 835 from 957 on abacus

Step-1: Place 957 on the abacus.
Step-2: Subtract 835 digit by digit start from left to right:
Hundreds: 9 - 8 = 1.
Tens: 5 - 3 = 2.
Units: 7 - 5 = 2.

The abacus should show 822.

Chapter – 3

Vedic Tricks of Number Addition

Addition is the most fundamental mathematical operation, serving as the basis for more complex calculations. It involves combining numbers to find their sum. A key concept in addition is that by repeatedly adding the number 1 to the previous result, starting from unity (the number 1), all the natural numbers can be generated.

This principle is encapsulated in the first Vedic sutra *"Ekadhiken Purvena"*, ("By One More than the One Before"), which describes the generation of numbers from unity. Thus, addition with the number 1 forms the cornerstone for counting and creating all other numbers.

Here are the examples of numbers generated by adding 1 to previous number –

$$
\begin{aligned}
0 \quad &+\,1 \quad &= 1 \\
1 \quad &+\,1 \quad &= 2 \\
2 \quad &+\,1 \quad &= 3 \\
3 \quad &+\,1 \quad &= 4 \\
4 \quad &+\,1 \quad &= 5 \\
5 \quad &+\,1 \quad &= 6 \\
6 \quad &+\,1 \quad &= 7 \\
7 \quad &+\,1 \quad &= 8 \\
8 \quad &+\,1 \quad &= 9 \\
9 \quad &+\,1 \quad &= 10 \text{ and so on.}
\end{aligned}
$$

In vedic mathematics, numbers are represented using the digits 0 to 9, which are considered the fundamental building blocks for numerical representation. Each digit is representing a unique value, and by combining these digits, can be express as higher digit numbers.

The digits 1 to 9 are considered most significant in vedic mathematics as they hold inherent qualities and properties. These digits, when combined in various ways, create a vast range of numbers. The digit 0 serves as a placeholder and holds positional value, allowing us to represent numbers in different places or orders of magnitude.

Through arithmetic operations like addition, subtraction, multiplication, and division, we can manipulate and work with these digits to perform calculations and solve mathematical problems. Vedic mathematics provides techniques and sutras that enable efficient mathematical calculations using these digits.

Find the Digit Sum (Using 9-Point Circle)

As we have already discussed that the number system uses just nine digits and zero. All greater numbers are just combinations of these. And this simple diagram has a great number of useful applications in calculation.

Let us find the digit sum of numbers making use of 9-point circle. Since all numbers, no matter how long it is, can be reduced to a single digit.

Example digit sum of $10 = 1 + 0 = 1$ and digit sum of $11 = 1 + 1 = 2$ and so on.

9-Point Circle –

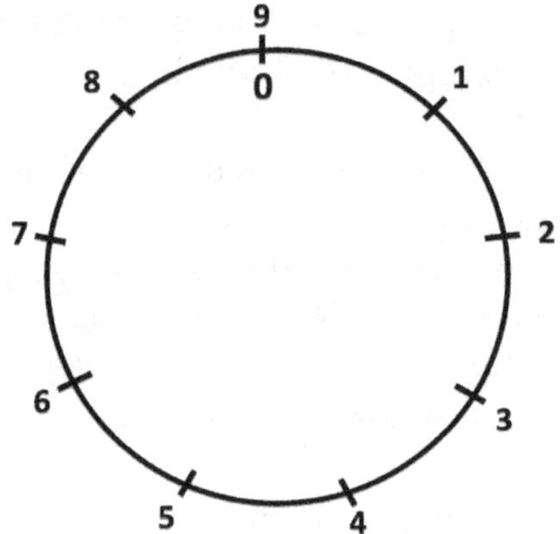

This circle shows the numbers from 1 to 9 and continuing to number round the circle would put 10 at the same place as 1.

And 11 would be at the same place as 2 and so on as shown below:

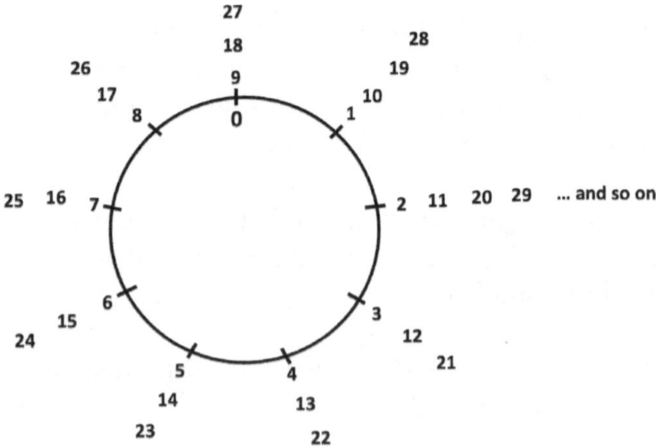

Notice also that 0 goes at the same place as 9 because, numbering backwards round the circle, 0 comes before 1. So all numbers will have their place somewhere round this circle.

Since, the digit sum of 10 is 1, therefore 10 is placed next to 1 on the circle. Similarly, the digit sum of 11 is 2, therefore 11 is placed next to 2 on the circle, and so on.

Let us try to find the digit sum of other numbers such as 12, 13, 14, 15,16, 17, 18, 19, 20, 21, 22, 23, 24, 25, 26, 27, 28, 29, …and so on and place them on the circle.

To find which branch of the circle a number is located, we simply find out its digit sum. Such as the digit sum of 23 is 5. Therefore we can place 23 on 5 branch of the circle. Similarly any bigger number can be placed on the suitable branch of the circle. For example, the digit sum of 88 = 8 + 8 = 7, So, 88 can be place on 7th branch of the circle.

Number Addition Completion or Non-completion (Using 10 Point Circle)

The sutra "Purana-puranabhyam" which means (*By the Completion or Non-completion*), shows the ability of wholeness.

10-Point Circle -

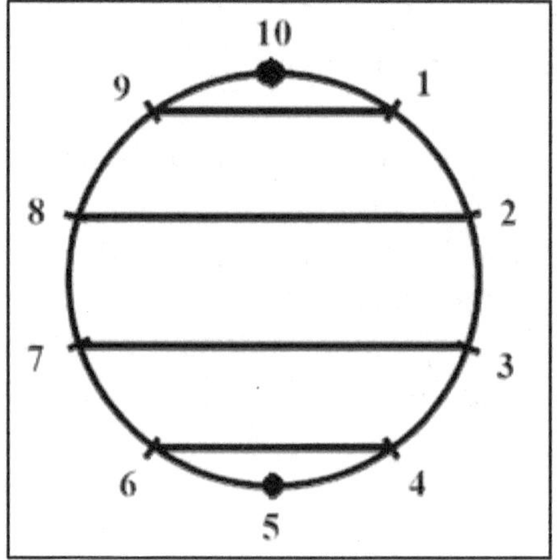

The representation of numbers from 1 to 10 on circle in known as 10-point circle. There are five unique pairs of numbers that add up to 10:

$$1 + 9 = 10 \qquad 2 + 8 = 10$$
$$3 + 7 = 10 \qquad 4 + 6 = 10$$
$$5 + 5 = 10$$

These pairs are very important in vedic addition of two numbers, where the last digit of two numbers add up to 10.

The 10 point circle can be used for adding on numbers, and for taking away, just as we use a number line.

Here, you can notice that all the numbers on any branch end with same last digit and all multiples of 10 appears on top branch. Here you will notice one interesting thing, whenever, we see the number pairs (which add to 10), then there is no need of manual addition.

Example-1: 38 + 22 ?

From 10-point circle we know 8 + 2 = 10
So we can write above sum as,
38 + 22 = 30 + 20 + 10 = **60 Ans.**

Example-1: 57 + 33 ?

From 10-point circle we know 7 + 3 = 10
So we can write above sum as,
57 + 33 = 50 + 30 + 10 = **90 Ans.**

Exercise do the following additions:

1) 19 + 11   **2)** 29 + 21   **3)** 39 + 31   **4)** 18 + 12
5) 28 + 22   **6)** 38 + 32   **7)** 17 + 13   **8)** 27 + 23
9) 37 + 33   **10)** 16 + 14 **11)** 26 + 24 **12)** 36 + 34
13) 15 + 15 **14)** 25 + 25 **15)** 35 + 35 **16)** 45 + 45

Additions by Making Multiple of 10, (Using Sutra *'By the Deficiency'*)

Which tell us how much a number differs from the wholeness.

9, 8 and 7 are close to 10
17, 18, 19, are close to 20
27, 28, 29, are close to 30
...
and so on

Example-1: 57 + 73 ?

57 is close to 60 and is 3 short.
73 is close to 70 and is 3 greater.

So, we can split number so as to make it multiple of ten
57 + 73 = 57 + (3 + 70) = 60 + 70 = 130 **Ans.**

Numbers Addition (Start Left to Right)

When we do addition with conventional method, we're often taught to start from the right and move left. But for mental math, it's actually better to work from left to right. This is because we read and say numbers from left to right, making it more natural to think about and calculate them in that order.

When we compute from right to left, we're essentially generating the answer backward, which can make mental math harder. Also, when estimating, it's more helpful to know the significant digits of the answer rather than just the last one. So, starting from the left helps us focus on the most important parts of the problem. While it might feel strange at first if you're used to working from right to left, with practice, you'll find it's the most efficient way to do mental calculations.

This left-to-right approach becomes especially crucial for larger addition, subtraction, multiplication, and division problems. Here are the examples of adding numbers from left to right -

Example-1
Add 73 and 32.

Here, we can solve the above addition with two basic methods

Method-1
    Add tenth place first
    i.e $(70 + 30) = $ **100**
    And then add unit place of numbers
    i.e. $(3+2) = $ **5**

Now, add both for final answer $100 + 5 =$ **105 Ans.**

Method-2
We can split 32 as **(30 +2)**
Now, first add **30** to 73, And then add **2**

i.e. 73 +32 = **(73 + 30) + 2 = 103 + 2 = 105 Ans.**

Example-2
Add the following numbers 134 + 321 + 413.

Steps: -

1. Add the leftmost digits (hundreds place value) in all numbers. Here, we have $1 + 3 + 4 = 8$ (i.e. 800).

2. Then, add the digits on the immediate right (tens place value) of all the given numbers. Here, we have $3 + 2 + 1 = 6$ (i.e. 60).

3. Then, we add the digits immediate right to it (units place value) of all the numbers. Here, we have $4 + 1 + 3 = 8$ (i.e., 8).

4. Now, add up all the final results (hundreds place + tens place + units place), will give the final result.

    $800 + 60 + 8 =$ **868 Ans.**

The Addition of Numbers, Where the Unit Digits of the Numbers Add Up to 10.

Here, we can use vedic sutra – "Purana-puranabhyam", which means (By the completion or non-completion). This sutra is used for addition problems, where the unit digits of the numbers add up to 10.

Example-1
To find-out the sum of the following numbers-
$$285 + 56 + 38 + 25 + 34 + 42 = ?$$

Steps:

Re-arrange the numbers in such a manner that their unit place add-up to 10

$$(285 + 25) \quad + \quad (56 + 34) \quad + \quad (38 + 42)$$
$$= \quad 310 \quad + \quad 90 \quad + \quad 80 = \mathbf{480}$$

Addition of List Number (Single Digit Pure Number)

In this method we start adding from bottom to top and drop the bar at the tens place and carry only the single digit forward, for addition of single digit number list, from 1 to 9.

<pre>
4   (5 + 4) = +9   7 ⁻   + 5   9 ⁻   + 1
7 ⁻   3   4 ⁻
6   2   1
7 ⁻   6 ⁻   5
5   7   7 ⁻
4 ⁻   2 ⁻   3
6   8   2
─────────   ─────────   ─────────
30+9 =39   30+5 =35   30+1 =31
</pre>

Steps: -

1) Start adding number from bottom, whenever, a sum is equal to ten or more, you need to drop a ten there, by putting a bar on number.

2) Add the remainder to the next number, if sum at the last number exceeds ten, then drop a bar on last number and kept in mind the remainder digit.

3) Now, count the number of bars, you drop (remember each bar indicates the number 10), and add the remainder digit to it, which will give the final result of addition.

Chapter – 4

Vedic Tricks of Subtraction

Concept of Complement and Base

In Vedic mathematics, the complement of a number refers to the difference between that number and a reference number, usually a power of 10. The complement is often used to simplify calculations.

To find the complement of a number, you subtract it from the nearest power of 10. For example, to find the complement of 47, you subtract 47 from 100 (the nearest power of 10), which gives you 53. So, the complement of 47 is 53.

Complements are particularly useful in subtraction problems. Instead of subtracting one number from another directly, you can take the complement of the number being subtracted and add it to the other number. This technique simplifies calculations and reduces the chance of errors.

In vedic mathematics, the base number is used as 10n format, actually base is nothing but, the power of 10 (i.e. 10, 100, 1000, 10000…and so on).

Find out Complement of Any Numbers From the Base.

The most useful vedic sutra *"Nikhilam Navatash Charamam Dashata"* (which means "All from 9 and last from 10"). The vedic sutra is used for instant subtraction of any number from the base, in other word the above sutra is used to calculate complement of that number instantly.

Example: To find out the complement of the number 0.749264 using Nikhilam Sutra.

The complement of given number means how much the number is less from the base
i.e.
Complement = (Base – Number)
Or
Complement = (1.0 - 0.749264)

Using vedic sutra "all from 9 and the last from 10".

9	9	9	9	9	10
- 7	4	9	2	6	4
2	5	0	7	3	6

Now, placing a decimal before six digits,
So, we can write above result as **0.250736.**

Learn Quick Calculation to Exchange While Shopping

Example-1:
If you buy a pencil and eraser at stationery shop for your kids of Rs. 8.53, if you pay rupee 10. What do you expect in return from seller?

Apply vedic sutra *"All from 9 and last from 10"*

```
      9   9   10
-     8.  5   3
      1.  4   7
```

So, you will receive **1.47** from seller, as balance.

Example-2:

Radha buy a pen for Rs.7.56 at stationery shop, for which she pay 20 rupee note. How much she will get in return as change from shopkeeper.

First of all you can split 20 rupee note into (10+10) rupee note to made mental calculation easy.

And then again apply sutra *"All from 9 and last from 10"*

```
      9   9   10
-     7.  5   6
      2.  4   4
```

Now, Add 10 to the above result (10 + 2.44 = 12.44), So, the Radha will have **12.44** back in return as balance.

Similarly, if you pay 50, 100, 200 and 500 rupee note to buy an item which cost below 50. You can apply the same tricks.

Complement of Numbers (0 to 9) From, Base 10.

Here, you can see the magic of combining number and its complementing, you will be able to write down the table of number 9.

Number	Complement	Table
0	9	= 09
1	8	= 18
2	7	= 27
3	6	= 36
4	5	= 45
5	4	= 54
6	3	= 63
7	2	= 72
8	1	= 81
9	0	= 90

Quickly Calculate Your Change, While Shopping!

Example-1

Rani buy a hair clip for Rs.5.67 at shop, for which she pay 50 rupee note. How much she will get in return as change.

Quick Mental Calculation

a) First of all deduct 10 rupee from 50, you will have 40 rupee.

b) Now find out complement of 5.67 from 10

Apply Vedic Sutra *"All from 9 and Last from 10"*

(9-5)   = **4**
(9-6)   = **3**
(10-7)   = **3**

We get 433, just place decimal after two digit, we got 4.33

So the remaining balance we will received back from shopkeeper will be (40 + 4.33), **i.e. 44.33 rupee**

Example-2

Shyam buy a geometry box for Rs.15.75 at stationery shop, for which he pay 100 rupee note. How much she will get in return as change.

Quick Mental Calculation

a) First of all deduct 20 rupee from 100, you will have 80 rupee.

b) Now find out complement of 5.75 from base 10

Apply Vedic Sutra *"All from 9 and Last from 10"*

(9-5)   = **4**
(9-7)   = **2**
(10-5) = **5**

You will have 4.25
So the remaining balance we will received back from shopkeeper will be (80 + 4.25), i.e. **84.25 rupee**

Example-3

Poonam buy a book for Rs.150.75 at stationery shop, for which he pay 200 rupee note. How much she will get in return as change.

Quick Mental Calculation

a) First of all deduct 160 rupee from 200, you will have 40 rupee.
b) Now find out complement of 0.75 from base 10

Apply Vedic Sutra *"All from 9 and Last from 10"*

(9-0)   = **9**
(9-7)   = **2**
(10-5) = **5**

You will have 9.25
So the remaining balance we will received back from shopkeeper will be (40 + 9.25), i.e. **49.25 rupee**

Chapter – 5

Vedic Tricks of Quick Multiplications

Multiplication of Number By 9

In Vedic mathematics, 9 is most interesting number, there are various techniques for multiplying numbers by 9. One of the commonly used methods is called "*Nikhilam Navatash Charamam Dashatah*" which translates to "*All from 9 and the last from 10*". Numbers may be multiply by 9, 99, 999, 9999….and so on, using the sub-sutra "*Ekanyunena Purvena*", means "By one less than the one before", followed by *Nikhilam*.

The multiplications by number 9 is divided into three parts –

A) *Where both multiplier and multiplicand have the same number of digits.*

Example-1
Multiply 8 × 9

i) By sub-sutra "One less than the One before"

We have
(8 -1) = 7
ii) By sutra "All from 9 and last from 10"

We have

```
  10
-  8
----
   2
```

So, from(i) and (ii) we have
$8 \times 9 = $ **7 2 Ans.**

Example-2
Multiply 356×999

i) By sub-sutra "One less than the One before"

We have
$(356 - 1) = $ **355**

ii) By sutra "All from 9 and last from 10"

We have

```
  9   9   10
- 3   5   6
-----------
  6   4   4
```

So, from(i) and (ii) we have
$356 \times 999 = $ **355 644** Ans.

B) Where multiplier have the more number of digits than the multiplicand.

Example-1
Multiply 32×999

In this situation first of all we make the digit equal, by placing zero before the multiplicand number.

So, we can write above multiplication as –
  032×999
Then follow the same trick to multiply -
i) By sub-sutra "One less than the One before"

By subtracting 1 from the given number, we have -
  $(032 - 1) = \mathbf{31}$

ii) By sutra *"All from 9 and last from 10"*

We have

9	9	10
- 0	3	2
9	6	8

So, from(i) and (ii) we have
$32 \times 999 = \mathbf{31\ 968}$ Ans.

C) *Where multiplier have the less number of digits than the multiplicand.*

Example-1
Multiply 342×99

In this case, we can divide the multiplicand number into two parts, so that the second part has the same number of digits as the multiplier.
We can divide 342 as -
   3: 42

i) Add '1' to the left part and subtract it from original number.
   $(3+1) = 4$
And then $(342 - 4) =$ **338** (First part of the result)

ii) Find the complement of the second part by sutra "All from 9 and last from 10".

9	10
- 4	2
5	8

(Second part of the result)

So, from(i) and (ii) we have,
$342 \times 99 =$ **338 58 Ans.**

Doubling & Halving Tricks

A) Multiplying and dividing numbers by 2, 4, 8...etc.

Doubling means (to make the number double) and halving means to make it half. It is advised to split bigger number into its place value first then apply the rule of doubling or halving accordingly, to find out its multiplication or division by numbers 2, 4, 8, 16….and so on.

Example-1
Multiply number 15 by 2, 4, 8, 16….and so on

By doubling method

15×2	= (double of 15)	= **30**
15×4	= (double of 30)	= **60**
15×8	= (double of 60)	=**120**
15×16	= (double of 120)	=**240**

Example-2
Divide number 64 by 2, 4, 8, 16…. and so, on

By halving method

$64 \div 2$	= (half of 64)	= **32**
$64 \div 4$	= (half of 32)	= **16**
$64 \div 8$	= (half of 16)	= **8**
$64 \div 16$	= (half of 8)	= **4**

B) Doubling or halving the numbers by splitting into its place value

Example-1
Doubling the number 44 by splitting method

By splitting into its place value, we can write 44 as
    $40 + 4$
Now by doubling
80 (double of 40) + **8** (double of 4) = **88**

Example-2
Halving the number 98 by splitting

By splitting
We can write 98 as
    $90 + 8$
Now by halving
45 (Half of 90) + **4** (Half of 8) = **49**

Example-3
Halving the number 3230 by splitting method

By splitting into its place value
We can write 3230 as
$3000 + 200 + 30$

Now by having
1500 (Half of 3000) + **100** (Half of 200) + **15** (Half of 30)
Now we have
1500 + 100 + 15 = 1615

C) Multiplying Any Numbers By 5 (Using Doubling and Halving Method)

Example-1
Multiply 5×68

By doubling
$$5 \times 68 \quad = (2 \times 5) \times 68$$
$$= 10 \times 68$$
$$= 680$$
Now, by halving
$$680/2 \quad = 340$$
Ans. 340

Example-2
Multiply 5×88

By doubling
$$5 \times 88 \quad = (2 \times 5) \times 88$$
$$= 10 \times 88$$
$$= 880$$
Now, by halving
$$880/2 \quad = 440$$
Ans. 440

Example-3
Multiply 5×206

By doubling
$$5 \times 206 = (2 \times 5) \times 206$$

$$= 10 \times 206$$
$$= 2060$$

Now, by halving

2060/2   = 1030 .

Ans. 1030

Example-4

Multiply 5 × 845

By doubling

$$5 \times 845 = (2 \times 5) \times 845$$
$$= 10 \times 845$$
$$= 8450$$

Now, by halving

8450/2   = 4225

Ans. 4225

Example-5

Multiply 5 × 380

By doubling

$$5 \times 380 = (2 \times 5) \times 380$$
$$= 10 \times 380$$
$$= 3800$$

Now, by halving

3800/2   = 1900

Ans. 1900

D) Multiplying Any Numbers By 50 (Using Doubling and Halving Method)

Example-1
Multiply 50 × 27

By doubling
50 × 27 = (2 x 50) × 27
= 100 × 27
= 2700
Now, by halving
2700/2 = 1350
= 1350 Ans.

Example-2
Multiply 50 × 387

By doubling
50 × 387 = (2 × 50) × 387
= 100 × 387
= 38700
Now, by halving
38700/2 = 19350
= 19350 Ans.

Example-3
Multiply 50 × 2652

By doubling
50 × 2652 = (2 × 50) × 2652
= 100 × 2652
= 265200

Now, by halving
265200/2   = 132600
        = 132600 Ans.
Example-4
Multiply 50 × 8020

By doubling
50 × 8020   = (2 × 50) × 8020
                = 100 × 8020
                = 802000
Now, by halving
802000/2   = 401000
        = 401000 Ans.

Example-5
Multiply 50 × 873

By doubling
50 × 873   = (2 × 50) × 873
            = 100 × 873
            = 87300
Now, by halving
87300/2 = 43650
        = 43650 Ans.

E) Multiplying Any Numbers By 25 (Using Doubling and Halving Method)

Example-1
Multiply 25 × 68

By doubling twice
25 × 68 = (2 × 2 × 25) × 68
= 100 × 68
= 6800
Now, by halving twice
6800/2 = 3400 and 3400/2 = 1700
= 1700 Ans.

Example-2
Multiply 25 × 108

By doubling twice
25 × 108   = (2 × 2 × 25) × 108
= 100 × 108
= 10800
Now, by halving twice
10800/2 = 5400 and 5400/2 = 2700
= Ans. 2700

Example-3
Multiply 25 × 34

By doubling twice
25 × 34 = (2 × 2 × 25) × 34

$$= 100 \times 34$$
$$= 3400$$

Now, by halving twice

$$3400/2 \quad = 1700 \quad \text{and } 1700/2 \quad = 850$$
$$\mathbf{= 850 \ Ans.}$$

Example-4

Multiply 25×56

By doubling twice

$$25 \times 56 \quad = (2 \times 2 \times 25) \times 56$$
$$= 100 \times 56$$
$$= 5600$$

Now, by halving twice

$$5600/2 \quad = 2800 \quad \text{and } 2800/2 \quad = 1400$$
$$\mathbf{= 1400 \ Ans.}$$

Example-5

Multiply 25×98

By doubling twice

$$25 \times 56 \quad = (2 \times 2 \times 25) \times 98$$
$$= 100 \times 98$$
$$= 9800$$

Now, by halving twice

$$9800/2 \quad = 4900 \quad \text{and } 4900/2 \quad = 2450$$
$$\mathbf{= 2450 \ Ans.}$$

F) Divide Any Numbers By 5 (Using Doubling and Divide by 10 Method)

Example-1

Divide $38 \div 5$

Doubling and then divide by 10

$= (2 \times 38) \div (2 \times 5)$

$= 76 \div 10$

$= 7.6$

$= \textbf{7.6 Ans.}$

Example-2

Divide $97 \div 5$

Doubling and then divide by 10

$= (2 \times 97) \div (2 \times 5)$

$= 194 \div 10$

$= 19.4$

$= \textbf{19.4 Ans.}$

Example-3

Divide $381 \div 5$

Doubling and then divide by 10

$= (2 \times 381) \div (2 \times 5)$

$= 762 \div 10$

$= 76.2$

$= \textbf{76.2 Ans.}$

Example-4
Divide $40 \div 5$

Doubling and then divide by 10
$= (2 \times 40) \div (2 \times 5)$
$= 80 \div 10 = 8$
= 8 Ans.

Example-5
Divide $1250 \div 5$

Doubling and then divide by 10
$= (2 \times 1250) \div (2 \times 5)$
$= 2500 \div 10$
$= 250$
= 250 Ans.

G) Divide Any Numbers By 50 (Using Doubling and Divide by 100 Method)

Example-1
Divide $37 \div 50$

Doubling and then divide by 100
$= (2 \times 37) \div (2 \times 50)$
$= 74 \div 100$
$= 0.74$
= 0.74 Ans.

Example-2
Divide $365 \div 50$

Doubling and then divide by 100
$= (2 \times 365) \div (2 \times 50)$
$= 730 \div 100$
$= 7.30$
= 7.3 Ans.

Example-3
Divide $825 \div 50$

Doubling and then divide by 100
$= (2 \times 825) \div (2 \times 50)$
$= 1650 \div 100$
$= 16.50$
= 16.5 Ans.

Example-4
Divide $125 \div 50$

Doubling and then divide by 100
$= (2 \times 125) \div (2 \times 50)$
$= 250 \div 100$
$= 2.50$
= 2.5 Ans.

Example-5
Divide $625 \div 50$

Doubling and then divide by 100
= (2 × 625) ÷ (2 × 50)
= 1250 ÷ 100
= 12.50
= 12.5 Ans.

H) Divide Any Numbers By 25 (Doubling Twice Then Divide by 100)

Example-1
Divide 725 ÷ 25

Doubling twice and then divide by 100
= (2 × 2 × 725) ÷ (2 × 2 × 25)
= 2900 ÷ 100
= 29.00
= 29 Ans.

Example-2
Divide 829 ÷ 25

Doubling twice and then divide by 100
= (2 × 2 × 829) ÷ (2 × 2 × 25)
= 3316 ÷ 100
= 33.16
= 33.16 Ans.

Example-3
Divide 25125 ÷ 25

Doubling twice and then divide by 100

$= (2 \times 2 \times 25125) \div (2 \times 2 \times 25)$

$= 100500 \div 100$

$= 1005.00$

$= 1005$ Ans.

Example-4

Divide $7525 \div 25$

Doubling twice and then divide by 100

$= (2 \times 2 \times 7525) \div (2 \times 2 \times 25)$

$= 30100 \div 100$

$= 301.00$

$= 301$ Ans.

Example-5

Divide $2525 \div 25$

Doubling twice and then divide by 100

$= (2 \times 2 \times 2525) \div (2 \times 2 \times 25)$

$= 10100 \div 100$

$= 101.00$

$= 101$ Ans.

Multiplications of Numbers with Patterns that Add to 10.

A) Squaring a Number, Ending with 5.

Example-1:
Find out the square of $(25)^2$

#Rule-1: *Remember, if last digit of a number is **5**, then last two digits of square will always be **25**.*

Mental steps:

Multiplying the first digit of number (2) with its next digit (3), so, we have

$\quad$ 2 × 3 = **6** (First part of the result)

Now, from *#Rule-1 the last digit of number is 5, so the last two digits of square will always be **25**.*

Therefore, second part of the result is **25**

Now, combining the two results, we have **6 25**
Thus, we, have $(25)^2 = $ **6 25**

Example-2:
Find out the square of $(105)^2$

Mental Steps:

Multiplying the first digit of number (10) with its next digit (11), so, we have

$\qquad 10 \times 11 = \mathbf{110}$ (First part of the result)

Now, from *#Rule-1 the last digit of number is 5, so the last two digits of square will always be **25**.*

Therefore, second part of the result is **25**

Now, combining the two results, we have **110 25**
Thus, we, have $(105)^2 = \mathbf{110\ 25}$

Example-3:
Find out the square of $(95)^2$

Mental Steps:

Multiplying the first digit of number (9) with its next digit (10), so, we have

$\qquad 9 \times 10 = \mathbf{90}$ (First part of the result)

Now, from *#Rule-1 the last digit of number is 5, so the last two digits of square will always be **25**.*

Therefore, second part of the result is **25**

Now, combining the two results, we have **90 25**
Thus, we, have $(95)^2 = \mathbf{90\ 25}$

Example-4:
Find out the square of (45)²

Mental Steps:

Multiplying the first digit of number (4) with its next digit (5), so, we have

$4 \times 5 = \mathbf{20}$ (First part of the result)

Now, from *#Rule-1 the last digit of number is 5, so the last two digits of square will always be 25.*

Therefore, second part of the result is **25**

Now, combining the two results, we have **20 25**
Thus, we, have $(45)^2 = \mathbf{20\ 25}$

Example-5:
Find out the square of (85)²

Mental Steps:

Multiplying the first digit of number (8) with its next digit (9), so, we have

$8 \times 9 = \mathbf{72}$ (First part of the result)

Now, from *#Rule-1 the last digit of number is 5, so the last two digits of square will always be 25.*

Therefore, second part of the result is **25**

Now, combining the two results, we have **72 25**
Thus, we, have $(85)^2 =$ **72 25**

Example-6:
Find out the square of $(115)^2$

Mental Steps:

Multiplying the first digit of number (11) with its next
digit (12), so, we have

    $11 \times 12 =$ **132** (First part of the result)

Now, from *#Rule-1 the last digit of number is 5, so the last
two digits of square will always be 25*.

Therefore, second part of the result is **25**

Now, combining the two results, we have **132 25**
Thus, we, have $(115)^2 =$ **132 25**

B) Multiplication of Numbers, Where "First Digits are Same, and Last Digits Add to 10"

Application of Vedic Sutra, "Antyayordasake'pi" which means "Last totalling to 10".

Example-1.
Find the multiplication 12 × 18 =?

Here, first digit of two numbers are same and sum of last digits are 10.

#Rule-1: Multiplying the first digit of number (n) with its next digit (n+1).

So, here we can multiplying the first digit of number (1) with its next digit (2).

$$1 × 2 = \mathbf{2} \text{ (First part of the result)}$$

Multiplying the last by last
$$2 × 8 \quad = \mathbf{16} \text{ (Second part of the result)}$$

Now, combining the two results, we have, 2 16
Thus, we, have 12 × 18 = **2 16 Ans.**

Example-2.
Find the multiplication 28 × 22 =?

First digit of two numbers are same and sum of last digits are 10.

#Rule-1: Multiplying the first digit of number (n) with its next digit (n+1).

So, here we can multiplying the first digit of number (2) with its next digit (3).

$2 \times 3 = \mathbf{6}$ (First part of the result)

Multiplying the last by last
$8 \times 2 \ = \mathbf{16}$ (Second part of the result)

Now, combining the two results, we have, 2 16
Thus, we, have $28 \times 22 = \mathbf{6\ 16\ Ans.}$

Example-3.
Find the multiplication $53 \times 57 =?$

First digit of two numbers are same and sum of last digits are 10.

#Rule-1: Multiplying the first digit of number (n) with its next digit (n+1).

So, here we can multiplying the first digit of number (5) with its next digit (6).

$5 \times 6 = \mathbf{30}$ (First part of the result)

Multiplying the last by last

$3 \times 7 \ = \mathbf{21}$ (Second part of the result)

Now, combining the two results, we have, 30 21
Thus, we, have 53 × 57 = **30 21 Ans.**

Example-4.
Find the multiplication 74 × 76 =?

First digit of two numbers are same and sum of last digits are 10.

#Rule-1: Multiplying the first digit of number (n) with its next digit (n+1).

So, here we can multiplying the first digit of number (7) with its next digit (8).

$7 \times 8 = \mathbf{56}$ (First part of the result)

Multiplying the last by last

$4 \times 6 \ = \mathbf{24}$ (Second part of the result)

Now, combining the two results, we have, 56 24
Thus, we, have 74 × 76 = **56 24 Ans.**

Example-5. Find the multiplication 84 × 86 ?

First digit of two numbers are same and sum of last digits are 10.

#Rule-1: Multiplying the first digit of number (n) with its next digit (n+1).

So, here we can multiplying the first digit of number (8) with its next digit (9).

$$8 \times 9 = \mathbf{72} \text{ (First part of the result)}$$

Multiplying the last by last
$$4 \times 6 = \mathbf{24} \text{ (Second part of the result)}$$

Now, combining the two results, we have, 72 24
Thus, we, have 84 × 86 = **72 24 Ans.**

Similarly, the following multiplication your-self using above tricks

1) **31 × 39** = 12 09
2) **13 × 17** = 02 21
3) **88 × 82** = 72 16
4) **44 × 46** = 20 24
5) **55 × 55** = 30 25
6) **67 × 63** = 42 21
7) **79 × 71** = 56 09
8) **32 × 38** = 12 16
9) **99 × 91** = 90 09
10) **34 × 36** = 12 24

C) Multiplication of Numbers, where "First Digits Add to 10" and Last Digits are Same".

Here we can see the application of Vedic Sutra *"Adhamadhenantya - manatayen" which means, "First by the first"* and *"Last by the last"*

Example-1.
Find the multiplication of 67 × 47 =?

#Rule-2: Multiply *"First by First"* and *Add the Last.*

Mental Steps:

(6 x 4) + 7   = **31** (First part of the result)

Multiplying "Last by the last"
       7 × 7   = **49** (Second part of the result)

Now, combining the two results, 67 × 47 = **3149 Ans.**

Example-2.
Find the multiplication 85 × 25 =?

#Rule-2: Multiply *"First by First"* and *Add the Last.*

Mental Steps:

(8 x 2) + 5   = **21** (First part of the result)

Multiplying "Last by the last"

$5 \times 5 \quad = \textbf{25}$ (Second part of the result)

Now, combining the two results, $85 \times 25 = \textbf{21 25 Ans.}$

Example-3.
Find the multiplication $38 \times 78=?$
#Rule-2: Multiply *"First by First"* and *Add the Last*.

Mental Steps:
$(3 \times 7) + 8 \quad = \textbf{29}$ (First part of the result)

Multiplying "Last by the last"
$8 \times 8 \quad = \textbf{64}$ (Second part of the result)
Now, combining the two results, $38 \times 78 = \textbf{29 64 Ans.}$

Example-4.
Find the multiplication $43 \times 63=?$

#Rule-2: Multiply *"First by First"* and *Add the Last*.

Mental Steps:

$(4 \times 6) + 3 \quad = \textbf{27}$ (First part of the result)

Multiplying "Last by the last"
$3 \times 3 \quad = \textbf{09}$ (Second part of the result)

Now, combining the two results, $43 \times 63 = \textbf{27 09 Ans.}$

D) Multiplication of Numbers, where the First Digits of the two Numbers are (n) and (n+1) and Last Digits Add to 10.

#Rule-3: Find the nearest base (multiple of 10), and split multiplication as (base - k) (base + k).

Example-1.
Find the multiplication of 36 × 44 ?

From *#Rule-3*, the nearest base is **40**
Now, split multiplication to (base - k) (base +k).

$$i.e.,\ 36 \times 44 = (40 - 4) \times (40 + 4)$$
$$= 40^2 - 4^2$$
$$= 1600 - 16$$
$$= 1584$$
$$= \textbf{1584 Ans.}$$

Example-2.
Find the multiplication 28 × 32 ?

From *#Rule-3*, the nearest base is **30**
Now, split multiplication to (base - k) (base +k).

$$i.e.,\ 28 \times 32 = (30 - 2) \times (30 + 2)$$
$$= 30^2 - 2^2$$
$$= 900 - 04$$
$$= 896$$
$$= \textbf{896 Ans.}$$

Example-3.

Find the multiplication 53 × 67 ?

From #*Rule-3*, the nearest base is **60**
Now, split multiplication to (base - k) (base +k).

i.e., 53 × 67 $= (60 - 7) \times (60 + 7)$
                  $= 60^2 - 7^2$
                  $= 3600 - 49$
                  $= 3551$
                  $= \mathbf{3551\ Ans.}$

Example-4.

Find the multiplication 74 × 86 ?

From #*Rule-3*, the nearest base is **80**
Now, split multiplication to (base - k) (base +k).

i.e., 74 × 86 $= (80 - 6) \times (80 + 6)$
                  $= 80^2 - 6^2$
                  $= 6400 - 36$
                  $= 6364$
                  $= \mathbf{6364\ Ans.}$

Example-5.

Find the multiplication 88 × 92 ?

From #*Rule-3*, the nearest base is **90**
Now, split multiplication to (base - k) (base +k).

$$i.e., 88 \times 92 = (90 - 2) \times (90 + 2)$$
$$= 90^2 - 2^2$$
$$= 8100 - 04$$
$$= 8096$$
$$= \textbf{8096 Ans.}$$

Example-6.
Find the multiplication 18×22 ?

From *#Rule-3*, the nearest base is **20**
Now, split multiplication to (base - k) (base +k).

$$i.e., 18 \times 22 = (20 - 2) \times (20 + 2)$$
$$= 20^2 - 2^2$$
$$= 400 - 04$$
$$= 396$$
$$= \textbf{396 Ans.}$$

Example-7.
Find the multiplication 12×28 ?

From *#Rule-3*, the nearest base is **20**
Now, split multiplication to (base - k) (base +k).

$$i.e., 12 \times 28 = (20 - 8) \times (20 + 8)$$
$$= 20^2 - 8^2$$
$$= 400 - 64$$
$$= 336$$
$$= \textbf{336 Ans.}$$

Example-8.
Find the multiplication 13 × 27?

From *#Rule-3*, the nearest base is **20**
Now, split multiplication to (base - k) (base +k).

$$
\begin{aligned}
i.e.\ 13 \times 27 \quad &= (20 - 7) \times (20 + 7) \\
&= 20^2 - 7^2 \\
&= 400 - 49 \\
&= 351 \\
&= \textbf{351 Ans.}
\end{aligned}
$$

Example-9.
Find the multiplication 14 × 26 ?

From *#Rule-3*, the nearest base is **20**
Now, split multiplication to (base - k) (base +k).

$$
\begin{aligned}
i.e.,\ 14 \times 26 \quad &= (20 - 6) \times (20 + 6) \\
&= 20^2 - 6^2 \\
&= 400 - 36 \\
&= 364 \\
&= \textbf{364 Ans.}
\end{aligned}
$$

Example-10.
Find the multiplication 15 × 25 ?

From *#Rule-3*, the nearest base is **20**
Now, split multiplication to (base - k) (base +k).

$$i.e., 15 \times 25 \quad = (20 - 5) \times (20 + 5)$$
$$= 20^2 - 5^2$$
$$= 400 - 25$$
$$= 375$$
$$= \textbf{375 Ans.}$$

Example-11.

Find the multiplication 16×24 ?

From *#Rule-3*, the nearest base is **20**
Now, split multiplication to (base - k) (base +k).

$$i.e., 15 \times 25 \quad = (20 - 4) \times (20 + 4)$$
$$= 20^2 - 4^2$$
$$= 400 - 16$$
$$= 375$$
$$= \textbf{384 Ans.}$$

E) Multiplication of Numbers, Where Last Digits of Both the Numbers is 5 and the Difference of Two Numbers are Equal to 10.

#Rule-1: *The last part of the answer is always 75*

#Rule-2: *Find out the square of bigger first digit, and minus 1 from the square (this will be first part of the answer).*

Example-1:

Multiply 45 × 35 ?

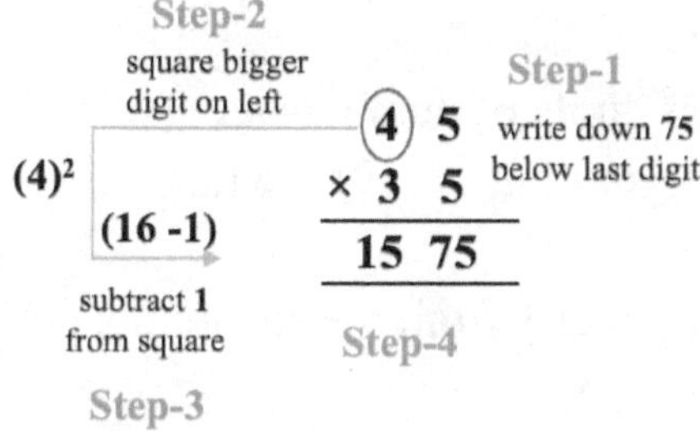

Follow the steps:

Step-1: From *#Rule-1*, write down **75** just below last digits (*the last part of the answer*)

Step-2: The first digits of numbers are 4 and 3, now find the square of 4 the bigger digit
i.e. $(4)^2 = 16$.

Step-3: Then subtract 1 from the square

i.e. (16-1) = 15.

Step-4: Write down the result 15 just below first digits (*the first part of the answer*).

Therefore, we get 45 × 35 = **15 75 Ans.**

Example-2:
Multiply 95 × 85 ?

Follow the steps:

Step-1: Write down **75** just below last digits of the numbers (the last part of the answer)

Step-2: The first digits of numbers are 9 and 8, now find the square of bigger digit i.e. $(9)^2 = 81$.

Step-3: Then subtract 1 from the square i.e. (81-1) = 80.

Step-4: Write down the result below first digits.

Step-2
square bigger
digit on left
$(9)^2$
(81 -1)
subtract 1
from square
Step-3

Step-1
9 5
× 8 5
80 75
write down 75
below last digit
Step-4

Therefore, we get 95 × 85 = **8075 Ans.**

Example-3:

Multiply 115×105 ?

Follow the steps:

Step-1: Write down **75** just below last digits of the numbers (the last part of the answer)

Step-2: The first digits of numbers are 11 and 10, now find the square of bigger digit i.e. $(11)^2 = 121$.

Step-3: Then subtract 1 from the square i.e. $(121-1) = 120$.

Step-4: Write down the result below first digits.

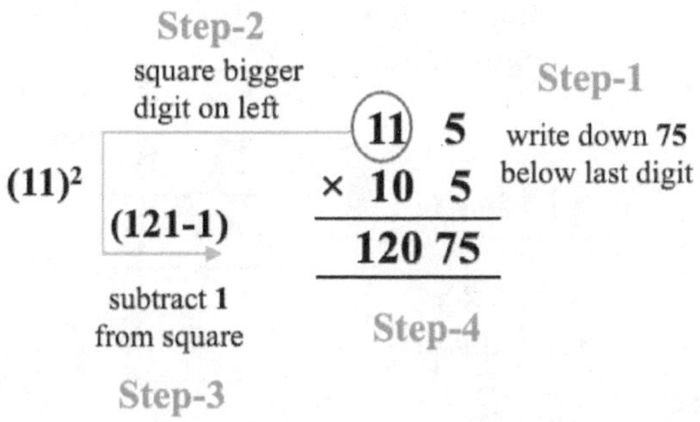

Therefore, we get $11 \times 10 = $ **12075 Ans.**

Example-4:

Multiply 125 × 115 ?

Follow the steps:

Step-1: Write down **75** just below last digits of the numbers (the last part of the answer)

Step-2: The first digits of numbers are 12 and 11, now find the square of bigger digit i.e. $(12)^2 = 144$.

Step-3: Then subtract 1 from the square i.e. $(144-1) = 143$.

Step-4: Write down the result below first digits.

Therefore, we get 12 × 11 = **14375 Ans.**

Multiplication of Any 2-Digit Numbers

Using Vedic Sutra "Urdhva-triyagbhyam", which means "Cross-wise and Vertical".

Here, is the graphic representation of sutra –

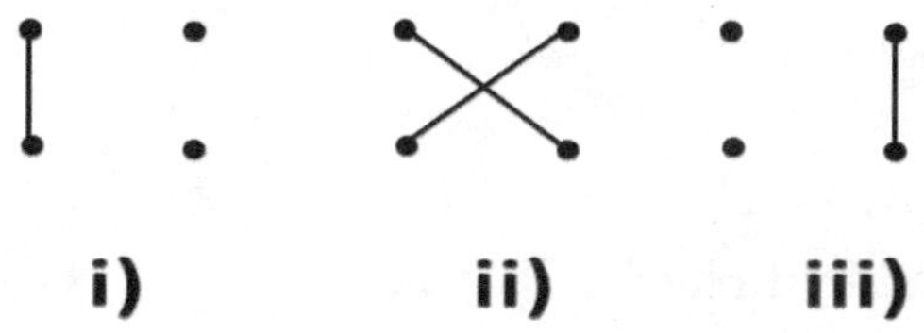

#Rule-1:
i) Vertical on left (*First by First*)
ii) Cris-Cross multiplication and Addition
iii) Vertical on right (*Last by Last*)

Example-1.
Find the following multiplications –

```
        3        4
    ×   2        5
___________________________
```

(3×2), $(2 \times 4 + 3 \times 5)$, (4×5)

Thus, we have, 6, $(8 + 15)$, 20

    6, 23, 20

Carry to be forward to the left

    6, 23, $_2$0 = 6, $_2$5, 0 = 8, 5, 0

Ans. 850

Multiplication of Numbers in Series of '1'

A) Multiplication of Two Numbers Ending with Last Digits "1"

Example-1.
Find the multiplications 81×31 ?

Easy mental steps:
    1) Multiply first by first $(8 \times 3) = $ **24**
    2) Add first by first $(8 + 3) = $ **11**
    3) Multiply last by last $1 \times 1 = 1$

We have, the results 24 11 1

Now, carry forward 1-digit to left (if sum is >9 in step-2)
So, we can write results as
$= 24_1\ 1\ 1$
$= 25\ 1\ 1$
Ans. 25 11

B) Multiplication method of two-digit numbers by 11.

#Rule-1: *Just place digit sum in middle of two digit, you will have the answer.*

Example -1.

Find the multiplications 23 × 11 ?

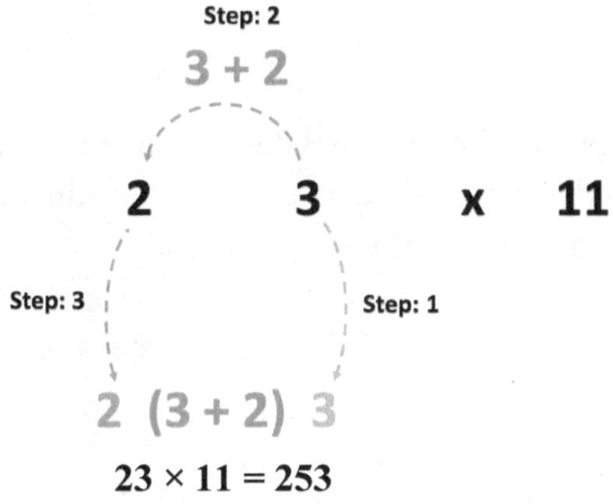

Mental Steps:

Steps-1:

Simply note down the last digit as it is. (..**3**)

Steps-2:

Now, add last digit to the left (3+2 = **5**)

Steps-3:

Then, write down the first digit as it is. (**2**..)
So, we can write the results

$$2 \quad 5 \quad 3$$

Thus, $23 \times 11 = \mathbf{2\,5\,3}$

Example-2.
Find the multiplications 35×11 ?
= 3 **(3+5)** 5
= 3 **8** 5
Ans. 385

The multiplication of any two-digit number by 11, takes only few seconds and method is much more quicker than the conventional method of multiplication.

Here, you can practice some more examples of similar kind, given below:

1) 11×11 : 1 (1+1) 1 = 121
2) 10×11 : 1 (0+1) 0 = 110
3) 22×11 : 2 (2+2) 2 = 242
4) 27×11 : 2 (7+2) 7 = 297
5) 33×11 : 3 (3+3) 3 = 363
6) 39×11 : 3 (9+3) 9 = 429
7) 44×11 : 4 (4+4) 4 = 484
8) 46×11 : 4 (6+4) 6 = 506
9) 51×11 : 5 (1+5) 1 = 561
10) 55×11 : 5 (5+5) 5 = 605
11) 66×11 : 6 (6+6) 6 = 726
12) 67×11 : 6 (7+6) 7 = 737
13) 68×11 : 6 (8+6) 8 = 748

14) **72** × **11** :   7 (2+7) 2   = 792
15) **77** × **11** :   7 (7+7) 7   = 847
16) **87** × **11** :   8 (7+8) 7   = 957
17) **88** × **11** :   8 (8+8) 8   = 968
18) **93** × **11** :   9 (3+9) 3   = 1023
19) **98** × **11** :   9 (8+9) 9   = 1078
20) **99** × **11** :   9 (9+9) 9   = 1089

Note: *if addition of the middle digits becomes tens place (double digit number), then carry would be forwarded to the next digit on the left for final result.*

C) Multiplication of 3-digit numbers by 11

#Rule-1: *Start adding pair of digits from left to right and place sum in middle of first and digit.*

Example -1
Find the multiplications 423×11 ?

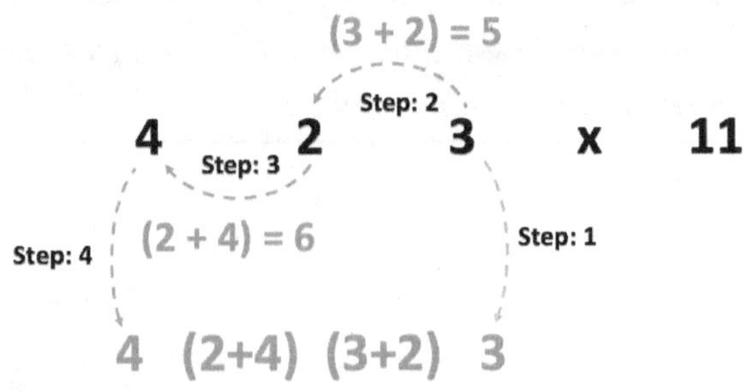

$$423 \times 11 = \textbf{4653 Ans.}$$

Mental Steps

Steps-1:
Simply note down the last digit as it is. (...**3**)

Steps-2:
Add last digit to the left (3+2 = **5**)

Steps-3:
Add second last digit to the left (2+4 = **6**)

Steps-4:
Then, write down the first digit as it is. (**4**...)

So, we can combine the results 4   6   5   3

Hence, 423 × 11 = **4653 Ans.**

Example - 2

Find the multiplications 234 x 11?

Steps:

    i) Start from last digit, simply note down the last digit as it is. (........**4**).

    ii) Now, add last digit to the left middle digit (4+3 = **7**).

    iii) Then, add middle digit to the left most digit (3+2 = **5**).

    iv) Then, write down the first digit as it is. (**2**.....).

Now, we can combine all the result, we have **2 5 7 4**

So, 234 × 11 = **2574 Ans.**

D) Multiply numbers with larger digits by 11.

#Rule-1: *We follow the same rule as for 3-digit multiplication by 11, i.e. start adding pair of digits from left to right and place sum in middle of first and digit.*

Example -1
Find the multiplications 3621342 × 11 ?

Follow the steps:
   i) Note down the last digit as it is : **2**
   ii) Add last digit to the left : 2 + 4 = **6**
   iii) And then, add second last digit to the left: 4+3 = **7**
   iv) Now add third last digit to the further left digit and continue the process to add until the first digit of the number has been reached -
   = 3+1 = **4**.
   = 1+2 = **3**.
   = 2+6 = **8**.
   = 6+3 = **9**.
v) The last step, write down the first digit as it is: **3**.

So, we can combined the result 3 9 8 3 4 7 6 2
Hence, we get 3621342 × 11 = **39834762 Ans.**

Here can have some more examples of multiplication by 11to practice your own with different numbers for better performance.

Practice the number multiplications by 11

i) 636×11 $= 6996$

ii) 706×11 $= 7766$

iii) 981×11 $= 10791$

iv) 546×11 $= 6006$

v) 360×11 $= 3960$

vi) 9010×11 $= 99110$

vii) 1234×11 $= 13574$

viii) 7020×11 $= 77220$

ix) 10187×11 $= 112057$

x) $312421 \times 11 = 3436631$

E) Multiplication of 2-digit numbers by triple digit "111-series".

Example -1.
Find the multiplications 43 × 111 ?

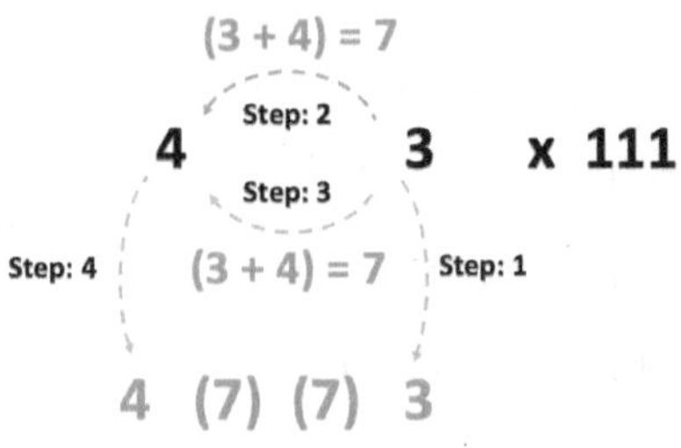

Thus, we get 43 × 111 = **4773 Ans.**

Here, are simple mental steps to perform the above multiplications.

Steps-1: Simply note down the last digit as it is. (...**3**)

Steps-2: Add last digit to the left (3+4 = **7**)

Steps-3: Add last digit to the left, one-more time (3+4 = **7**)

Steps-4: Then, write down the first digit as it is. (**4**...)

we can combine the results 4 7 7 3
Hence, we get 43 × 111 = **4773 Ans.**

F) Multiplication of 3-digit numbers by triple digit "111-series".

Example -1

Find the following multiplications 423×111

Here, are simple mental steps to perform the above multiplications.

Steps- 1: Simply note down the last digit as it is **3**

Steps-2 : Add last digit to next digit on the left **(3+2) = 5**

Steps-3 : Add three-digits start from last **(3+2+4) = 9**

Steps-4 : Add middle digit to next on the left **(2+4) = 6**

Steps-5 : Write down the first digit as it is **4**

we can combine the results 4 6 9 5 3

Hence, we get $423 \times 111 = $ **46953 Ans.**

G) Multiplication of 4-digit numbers by triple digit "111-series".

Example -1
Find the multiplications of 4253 × 111?

Here, are simple mental steps to perform the above multiplications-

Steps- 1:
Simply note down the last digit as it is **3**

Steps-2:
Add last digit to next digit on the left **(3 + 5) = 8**

Steps-3 :
Add three-digits start from last **(3 + 5 + 2) = 10**

Steps-4 :
Add three-digits start from second last **(5 + 2 + 4) = 11**

Steps-5 :
Add middle digit to next on the left **(2 + 4) = 6**

Steps-6 :
Write down the first digit as it is **4**
So, we can combine the result as **4 6₁ 1₁ 0 8 3**
Hence, we get 4253 × 111 = **472083 Ans.**

H) Multiplication of 5-digit numbers by triple digit "111-series".

Example -1: Find the multiplications of 43123 × 111 ?

Here, are simple mental steps to perform the above multiplications-

Steps- 1: Simply note down the last digit as it is **3**

Steps-2: Add last digit to next digit on the left **(3+2) = 5**

Steps-3: Add three-digits start from last **(3+2+1) = 6**

Steps-4: Add three-digits start from second last **(2+1+3) = 6**

Steps-5: Add three-digits start from third last **(1+3+4) = 8**

Steps-6: Add two-digits start from fourth digit to left **(3+4) = 7**

Steps-7: Write down the first digit as it is **4**

So, we can write the result 4 7 8 6 6 5 3
Hence, we get 43123 × 111 = **4786653 Ans.**

Multiplying Numbers From 12 to 19

Let us assume given multiplication (***ab*** × ***ac)***, where, ***a*** *is tenth digit and* ***b*** *is unit digit of first number, similarly* ***c*** *is unit digit of the second number.*

#Rules-1:
Add last digit c to first number
$$(ab + c)$$

#Rules-2:
Multiply the result by 10
$$(ab + c) \times 10$$

#Rules-3:
Now, multiply *Last by Last*
$$(b \times c)$$

#Rules-4:
Add the two results, we will get the final answer.

Example 1

Find the multiplications of 13 × 19 ?

From #Rules-1: Add last digit **9** to the first number **13**
(13 + 9) = **22**

From #Rules-2:
Multiply the result by 10
(22) × **10 = 220**

From #Rules-3:
Multiply *Last by Last*
(3 × 9) = **27**

From #Rules-4:
Add two results
220 + 27 = 247
Ans. 247

Similarly, we can find the following multiplications

i) **12 × 13 ?**
(12 + 3) 10 = 150
(2 × 3) = 6
150 + 6 = 156
Ans. 156

ii) **14 × 17 =?**
(14 + 7) 10 = 210
(4 × 7) = 28

$210 + 28 = 238$

Ans. 238

iii) 12 × 15 ?

$(12 + 5) \, 10 = 170$

$(2 \times 5) = 10$

$170 + 10 = 180$

Ans. 180

iv) 16 × 13 ?

$(16 + 3) \, 10 = 190$

$(6 \times 3) = 18$

$190 + 18 = 208$

Ans. 208

v) 17 × 18 ?

$(17 + 8) \, 10 = 250$

$(7 \times 8) = 56$

$250 + 56 = 306$

Ans. 306

vi) 12 × 18 ?

$(12 + 8) \, 10 = 200$

$(2 \times 8) = 16$

$200 + 16 = 216$

Ans. 216

Multiplication of Numbers by 12

By sutra "Sopantyadvaya- mantyam" (means "the ultimate and twice the penultimate").

#Rules-1:
First of all make a number zero sandwich by placing zero on either side of a given number.
   0 *number* **0**

#Rules-2:
By sutra "the ultimate and twice the penultimate", keep adding the last digit and twice the second last.

Example-1: Multiply the number 3542 × 12 ?

#Rules-1: First make a number zero sandwich by placing zero on either side of a given number.
0 3542 **0**

#Rules-2: By sutra "the ultimate and twice the penultimate", keep adding the last digit and twice the second last.

$$0 + 2*(2) = 4$$
$$2 + 2*(4) = 10$$
$$4 + 2*(5) = 14$$
$$5 + 2*(3) = 11$$
$$3 + 2*(0) = 3$$

We, can note down the results as reading from bottom to top -

   3 11 14 10 4

Carry forward '1' to the left digit

   $3_11_14_104$

   4 2 5 0 4

Therefore, we have $3542 \times 12 = \mathbf{42504}$

Ans. 42504

Example-2: Multiply 412353×12 ?

#Rules-1:

Make a zero sandwich

0 412353 **0**

#Rules-2:

By sutra "the ultimate and twice the penultimate", keep adding the last digit and twice the second last

   $0 + 2*(3) = \mathbf{6}$
   $3 + 2*(5) = \mathbf{13}$
   $5 + 2*(3) = \mathbf{11}$
   $3 + 2*(2) = \mathbf{7}$
   $2 + 2*(1) = \mathbf{4}$
   $1 + 2*(4) = \mathbf{9}$
   $4 + 2*(0) = \mathbf{4}$

Write down the above results start from bottom to top
    4 9 4 7 11 13 6

Carry forward '1' to left for every double-digit and write down the final result
    4 9 4 7_11_13 6
    4948236
So, we get 412353 x 12 = **4948236 Ans.**

Practice the following multiplications your own.

i)   593324 × 12

ii)   254578 × 12

ii)   912927 × 12

iv)   895713 × 12

v)   635717 × 12

vi)   295387 × 12

vii)   356128 × 12

viii)   495843 × 12

ix)   295729 × 12

x)   535135 × 12

xi)   595748 × 12

xii)   811571 × 12

Base Multiplications of Numbers

A) Multiplication by the numbers which is "One more than the Base"

#Rules-1:
Just repeat the given number twice

Example-1
Find the multiplications of 27×101 ?

Here base is 100
So, we just repeat the given number 27 twice to get answer

$27 \times 101 = 27\ 27$ Ans.

Explanation
$\quad 27 \times 101$
$\quad 27 \times (100 + 1)$
$\quad = 2700 + 27$
$\quad = 2727$
$\quad$ **Ans. 27 27**

Example-2
Find the multiplications of 349×1001 ?
Here base is 1000
So, we just repeat the given number 349 twice to get answer

$349 \times 1001 = 349\ 349$Ans.

Explanation

$349 \times (1000 + 1)$

$= 349000 + 349$

$= 349\ 349$

Ans. 349 349

Example-3

Find the multiplications of 53×1001 ?

Ans. 53 0 53

Example-4

Find the multiplications of 324×101 ?

The answer should be of 5-digit so 3 will be carry forward and add to left digit 7.

$324 \times 101 = 324\ 324$

$= 324_324$

$= 32724$

Ans. 32724

Example-5

278×101

$= 278\ 278$

$= 278_278$

$= 28078$

Ans. 28078

Example-6

568×1001

$= 568\ 568$

Ans. 568 568

Example-7

273×1001

$= 273\ 273$

Ans. 273 273

Example-8

970×101

$= 970\ 970$

$= 970_9 70$

$= 97970$

Ans. 97970

Example-9

5012×10001

$= 5012\ 5012$

Ans. 5012 5012

Example-10

601×10001

$= 601\ 0\ 601$

Ans. 601 0 601

B) Proportionate multiplication of number to the given floating base.

Example:

The proportionate floating base for number 102030 is 100000, 2000, and 30 respectively. This kind of multiplication can be performed by Vedic Sutra "Anurupyena", which means (proportionately). Suppose we multiply number *ab* by 10203040506

#Rule-1: Remove all the zeros, and replace non-zero digits 1, 2,3,4,5, and 6 by number *ab*.

#Rule-2:
Multiply the new number by 1, 2, 3, 4, 5, and 6 proportionately, we will get final answer.

Example-1
1) 35 × 101   = **35 35**
2) 27 × 201   = **54 27**
3) 15 × 301   = **45 15**
4) 30 × 401   = **120 30**

Example-2
1) 35 × 10203   = **3570105**
2) 8 × 1020304   = **8 16 24 32**
3) 5 × 504030201   = **25 20 15 10 05**
4) 12 × 501   = **60 12**

C) Multiply by Splitting Number Into Smaller Factors (Using Vedic Sutra 'Proportionately').

Example-1

1) $63 \times 67 \quad = (3 \times 21) \times 67$
$\qquad = 21 \times (3 \times 67)$
$\qquad = 21 \times 201$
$\qquad = \textbf{42 21 Ans.}$

2) $93 \times 67 \quad = (3 \times 31) \times 67$
$\qquad = 31 \times (3 \times 67)$
$\qquad = 31 \times 201$
$\qquad = \textbf{62 31 Ans.}$

3) $50 \times 55 \quad = (2 \times 25) \times 55$
$\qquad = 25 \times (2 \times 55)$
$\qquad = 25 \times 110$
$\qquad = 25\ 25\ 0$
$\qquad = 25_2\ 5\ 0$
$\qquad = \textbf{27 5 0 Ans.}$

4) $24 \times 37 \quad = (3 \times 8) \times 37$
$\qquad = 8 \times (3 \times 37)$
$\qquad = 8 \times 111$
$\qquad = \textbf{8 8 8 Ans.}$

D) Multiplication of Number with the help of Base and Working Base.

Base is the number which can be written in the power 10. (i.e. 10, 10^2, 10^3, 10^4, 10^5.....etc. or 10, 100, 10000, 100000.....etc.), whereas the working base is near to the actual number.

Number	Base	Working Base
43	10	40
97	10	100
212	100	200
402	100	400
517	100	500
2015	1000	2000
3019	1000	3000

Example-1:
Multiply 53×56 ?

Follow the steps:

Step-1:
Write down base and working base on top

Step-2:

Find the deviations of numbers from working base and then multiply the deviations i.e. (+3) × (+6) =**18.**

Step-3:

Add deviation of number to the other number on the cross i.e. (53+6) = **59.**

Step-4:

Write down the working base **50** on the left.

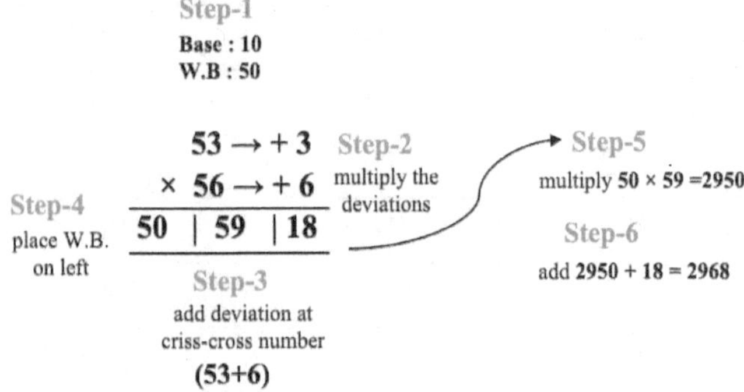

Step-5:

Multiply working base to the middle result i.e. (50 × 59) =**2950.**

Step-6:

Now add 18 to 2950,
So, we get (2950 + 18) = **2968 Ans.**

Multiplication of 3-digit Numbers by Criss-Cross Trick

By sutra "urdhva-triyagbhyam" (which means, vertically and crosswise). The graphical representation of the sutra –

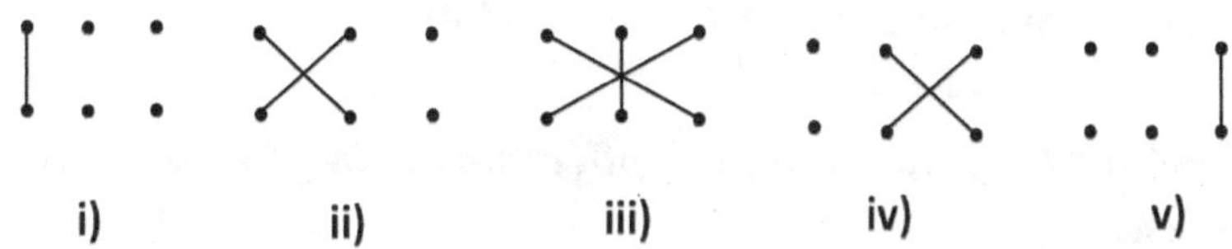

i) ii) iii) iv) v)

Example-1 Find the multiplications 342×615?

By following the above graphical steps:
i) Vertically on left
$(6 \times 3) = 18$

ii) Crosswise on left
$(6 \times 4) + (1 \times 3) = 27$

iii) Crosswise on extreme and mid-vertically
$(6 \times 2) + (1 \times 4) + (5 \times 3) = 31$

iv) Crosswise on right
$(1 \times 2) + (5 \times 4) = 22$

v) Vertically on right
$(5 \times 2) = 10$

Combining all the results
18, 27, 31, 22, 10

Carry forward left,
So, we get $= 18_2\ 7_3\ 1_2\ 2_1\ 0\quad = 2\ 0_1\ 0\ 3\ 3\ 0$

Thus, we have, $342 \times 615 =$ **210330 Ans.**

Practice the following multiplications using Criss-Cross Trick

i)	485×693
ii)	928×233
iii)	727×853
iv)	423×213
v)	551×203
vi)	728×536
vii)	992×235
viii)	828×123
ix)	425×234
x)	111×111
xi)	222×222
xii)	333×333

Chapter – 6

Quick Technique of Squaring Numbers

Special Tricks for Squaring Any Numbers

As you know that a square of number is nothing but multiplication of number itself -

#Rule: Split two digit number *(mn)* into *(a+b)* and then apply formula $(a+b)^2 = a^2 + b^2 + 2ab$.

Thus, we can write the following relation
$(mn)^2 = (a+b)^2 = a^2 + b^2 + 2ab$

Example-1
Find Square of numbers $(53)^2$?

$(mn)^2 = (a+b)^2 = a^2 + b^2 + 2ab$

$$(53)^2 = (50+3)^2 = (50)^2 + (3)^2 + 2(50 \times 3)$$
$$= 2500 + 9 + 2(150)$$
$$= 2500 + 9 + 300$$
$$= \mathbf{2809 \ Ans.}$$

Short Tricks of Finding Square of $(53)^2$?

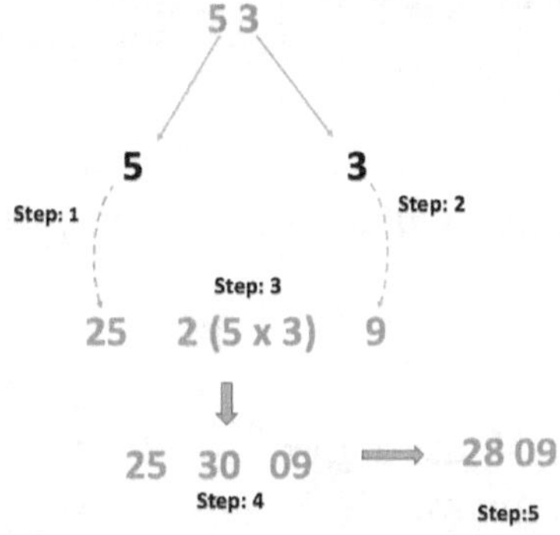

Step-1:
Find square of first digit
$$(5)^2 = \textbf{25}$$

Step-2:
Find square of second digit
$$(3)^2 = \textbf{9}$$

Step-3:
Multiply two both digits to each other and double

$2 \times$ (first digit $\times$ second digit),

Then place the result in the middle. i.e., $2 (5 \times 3) = \textbf{30}$

So, we get, 25 30 9

Since the result of square should be less than four digit. So, we can carry forward **3** to left.

Therefore, the above result can be written as **2809 Ans.**

Squaring of Numbers From 50 to 59

Let's assuming square of $(ab)^2$

#Rule-1: Add **25** to the unit digit (**b)**
i.e. $(25 + b)$
This is first part of the answer

#Rule-2: Find out the square of last digit b^2
This is the second part of the answer

#Rule-3: If square of b^2 < the double digit, then place zero before it. (e.g. for $b = 3$, we have $3^2 = 3 \times 3 = 09$)

Let's understand the above tricks with examples

Example-1
Find the square of $(50)^2$

Here, are simple mental steps

#Rule-1: Add **25** to **b**
$(25 + 0) = 25$
This will be the first part of the answer

#Rule-2: Find out the square of last digit b^2 (which give the second part of the answer)
$b^2 = 0^2 = 00$
Combining two results, we have 25 00
Ans. 25 00.

Example-2
Find the square of $(51)^2$

We can easily find out the square of number from *#Rule-1* and *#Rule-2*

$(25 + 1) = 26$
$(1 \times 1) \quad = 01$
Ans 26 01.

Example-3
Find the square of $(52)^2$

$(25 + 2) = 27$
$(2 \times 2) \quad = 04$
Ans 27 04.

Example-4
Find the square of $(53)^2$

$(25 + 3) = 28$
$(3 \times 3) \quad = 09$
Ans 28 09.

Example-5
Find the square of $(54)^2$

$(25 + 4) = 29$
$(4 \times 4) \quad = 16$
Ans 29 16.

Example-6
Find the square of $(55)^2$

$(25 + 5) = 30$
$(5 \times 5) \quad = 25$
Ans 30 25.

Example-7
Find the square of $(56)^2$

$(25 + 6) = 31$
$(6 \times 6) \quad = 36$
Ans 31 36.

Example-8
Find the square of $(57)^2$

$(25 + 7) = 32$
$(7 \times 7) \quad = 49$
Ans 32 49.

Example-9
Find the square of $(58)^2$

$(25 + 8) = 33$
$(8 \times 8) \quad = 64$
Ans 33 64.

Example-10
Find the square of $(59)^2$

$(25 + 9) = 34$
$(9 \times 9) \quad = 81$
Ans 34 81.

Squaring the Numbers, which are Near to Base "By the Deficiency" Tricks

Squaring the numbers close to base 10, 100, 1000 and so on, here we can apply vedic sutra *"Yavadunam"*, which means (*By the deficiency*).

#Rule-1: Find out the deficiency or surplus ($\pm d$) value from the base

#Rule-2: Add the difference to a given number ($n \pm d$) (this will give the first part of answer)

#Rule-3: Find square of deficiency or surplus ($\pm d$)^2
This will give the second part of answer

Let's understand the tricks with examples

Example-1.
Find the square of $(13)^2 = ?$

#Rule-1: Find out the deficiency or surplus value from the base.
Here, base is **10** and ($\pm d$) value is **3**

#Rule-2: Add the difference to a given number ($n \pm d$)

Now, have $(13 + 3) = $ **16** (*Ans. of First Part*)

#Rule-3: Find square of ($\pm d$)^2
Square of $(3)^2$ is 9 (*Ans. of Second Part*)

Now combining the two results
$(13)^2 = $ **16 9 Ans.**

Example-2.
Find the square of $(104)^2 = ?$

#Rule-1: Find out the deficiency or surplus value from the base.
Here, base is **100** and deficiency or surplus value is **4**

#Rule-2: Add the difference to a given number (this will give the first part of answer)

Now, have $(104 + 4) = $ **108**

#Rule-3: Find square of deficiency or surplus value (this will give the second part of answer)

Square of $(4)^2$ is **16**

Now, combining the two results
$(104)^2 = $ **108 16 Ans.**

Example-3.
Find the square of $(97)^2 = ?$

#Rule-1: Find out the deficiency or surplus value from the base.
Here, base is **100** and deficiency or surplus value is **3**

#Rule-2: Add the difference to a given number (this will give the first part of answer)

Now, have $(97 + (-3)) = $ **94**
Note: *Deficiency is added with negative sign.*

#Rule-3: Find square of deficiency or surplus value (this will give the second part of answer)

Square of $(3)^2$ is **09**

Now, combining the two results
$(97)^2 =$ **94 09 Ans.**

Example-4.
Find the square of $(1007)^2 =$?

#Rule-1: Find out the deficiency or surplus value from the base.
Here, base is **1000** and deficiency or surplus value is **7**

#Rule-2: Add the difference to a given number (this will give the first part of answer)

Now, have $(1007 + 7) =$ **1014**

#Rule-3: Find square of deficiency or surplus value (this will give the second part of answer)

Square of $(7)^2$ is **049**

Now, combining the two results
$(1007)^2 =$ **1014 049 Ans.**

Squaring a numbers with repeated n-digits

In general, squaring a number with n repeated digits, its square follows a palindrome pattern with the sequence of numbers from 1 to n and then back to 1.

The result reflects a mirrored sequence from 1 to n and then back to 1, creating a visually appealing structure. This phenomenon is rooted in the inherent symmetry of the square of a number, showcasing a delightful arrangement of digits that is both mathematical and aesthetically pleasing.

Example-1:

Squaring a numbers with repeated digits ($n = 1$)

$$(1)^2 = 1$$
$$(11)^2 = 121$$
$$(111)^2 = 12321$$
$$(1111)^2 = 1234321$$
$$(11111)^2 = 123454321$$
$$(111111)^2 = 12345654321$$
$$(1111111)^2 = 1234567654321$$
$$(11111111)^2 = 123456787654321$$
$$(111111111)^2 = 12345678987654321$$

Here, you can see a palindrome pattern in the result of square with the sequence of numbers from 1 to n and then back to 1.

Example-2:
Squaring a numbers with repeated digits ($n = 2$)

#Rule-1: Take $(n)^2$ as common, then follows a palindrome pattern with the sequence of numbers from 1 to n and then back from n to 1, then multiply by $(n)^2$.

Here, are the Examples -

$$(2)^2 \qquad = (2)^2 (1)^2 \quad = 4(1) \qquad\qquad = 4$$
$$(22)^2 \qquad = (2)^2 (11)^2 \ = 4(121) \qquad = 484$$
$$(222)^2 \qquad = (2)^2 (111)^2 = 4(12321) \qquad = 49284$$
$$(2222)^2 \quad = (2)^2 (1111)^2 = 4(1234321) \quad = 4937284$$
$$(22222)^2 \ = (2)^2 (11111)^2 = 4(123454321) = 493817284$$

Example-3:
Squaring a numbers with repeated digits ($n = k$)

#Rule-1: Take $(k)^2$ as common, then follows a palindrome pattern with the sequence of numbers from 1 to n and then back from n to 1, then multiply by $(k)^2$.

Here, are the Examples -

$$(k)^2 \qquad\qquad = (k)^2 (1)^2 \qquad = (k)^2 (1)$$
$$(k\,k)^2 \qquad\qquad = (k)^2 (11)^2 \qquad = (k)^2 (121)$$
$$(k\,k\,k)^2 \qquad\quad = (k)^2 (111)^2 \qquad = (k)^2 (12321)$$
$$(k\,k\,k\,k)^2 \qquad = (k)^2 (1111)^2 \qquad = (k)^2 (1234321)$$
$$(k\,k\,k\,k\,k)^2 \quad = (k)^2 (11111)^2 \quad = (k)^2 (123454321)$$

Chapter – 7

Vedic Techniques of Divisions

Division is a process of splitting a number into equal parts and finding out how many equal parts can be made. Which can be achieve by repeated subtraction from an original number.

We will deals with the divisions, where divisor are closer to the base. The base numbers are nothing but the number of (10^n) formats, where n = 1, 2, 3,… are power of base, e.g. 10, 100, 1000, 10 000…. etc., are the base numbers.

Now the another important part of number is its' complement, which can be obtain by simple vedic sutra "all from 9 and last from 10". That means a complement of a number is nothing but the difference from the base number.

Suppose, if you have a number *(n)* and *base (b)* then *complement (C)* can be obtain by following relationship

$$C(n) = (b - n)$$

Example -1.
Find the complement of 6, 7, 8 and 9?

Here, we can take base (b) = 10

So, can find the complements of given numbers -
C (6): (10 − 6) = **4**
C (7): (10 − 7) = **3**

C (8): (10 − 8) = **2**
C (9): (10 − 9) = **1**

Example – 2.

Find the complement of number 35678?

To find-out the complement of big number (or a number with larger digits) you need not established any base.

Here, you can directly find out the complement by using very important Vedic Sutra *"all from 9 and last from 10"*, which means subtract all digits from 9 and last digit from 10.

Start from left to right

```
    9   9   9   9   10
-   3   5   6   7    8
    7   4   3   2    2   (Complement)
```

Nikhilam Method of Division by 9.

Example-1:

Divide 71 by 9

```
9 | 7 | 1
1 | ↓ | 7
    | 7 | 8
```

Step-1:
Here, we will take base as 10, because 9 is near to it.

Step-2:
Find out the complement of 9, which is equal to 1.

The complement of 9 can be calculated as -
C (9) = (10 - 9) = 1

Step-3:
Drop the first digit 7 as it is (as indicated by arrow), and multiply with the complement 1 on left column and place in next column then add.

Step-4:
Check if, remainder is less than the divisor, here, we can see in above example remainder 8 is less than the divisor 9. So this can be declared as final remainder.
Thus, Quotient is **7** and Remainder is **8**.

Example-2:
Divide 356 by 9

9	3 5	6
1	↓ 3	8
	3 8	14

Since, remainder 14 is greater than 9, therefore, 1 will be added the quotient part to the left and 9 will be deducted from the 14. So, new quotient (38+1 = 39) and remainder (14 - 9 = 5)
Thus, Quotient: 39 and Remainder: 5

Generalized Method of Division by 9.

Let us first start with one of the special case of division (division by 9), it is a very interesting and simple technique. When you divide a number by 9, the remainder is always the digit sum of the original number.

2-Digit number divided by 9
   If $a\,b \div 9$

   Then
   Quotient = a
   Remainder = $(a + b)$

   In case where, remainder $(a + b) > 9$
   We have
   Quotient = $(a + 1)$
   Remainder = $(a + b) - 9$

Examples:
1) 13 divided by 9
   Here,
   Quotient is **1** and
   Remainder is (1+3) = **4.**

2) 14 divided by 9
   Here,
   Quotient is **1** and
   Remainder is (1+4) = **5.**

3) 17 divided by 9
   Here,
   Quotient is **1** and
   Remainder is (1+7) = **8.**

4) 41 divided by 9
Here,
Quotient is **4** and
Remainder is (4+1) = **5.**

5) 61 divided by 9
Here,
Quotient is **6** and
Remainder is (6+1) = **7.**

Note: *If remainder is greater than the 9, then we add 1 to the quotient and 9 is deducted from the remainder.*

---

6) 88 divided by 9
Here,
Quotient is **8** and
Remainder is (8+8) = **16** (*which is greater than 9*).
Therefore, we add 1 to quotient (i.e. 8+1 = **9**) and 9 is deducted from the remainder (i.e. 16-9 = **7**)
*So, new quotient is **9** and remainder is **7.***

3-Digit number divided by 9 (General Tricks)

If   a b c $\div$ 9
Then
Quotient = $ab + a$
Remainder = $(a + b + c)$

In case where, remainder $(a + b + c) > 9$

Quotient = $(ab + a) + 1$
Remainder = $(a + b + c) - 9$

Examples
1) 164 divided by 9

Here,
Quotient = *(ab +a)* = (16 + 1) = **17**
Remainder = (1 + 6 + 4) = **11** (> 9).

Since, remainder is greater than 9, so, we can add 1 to quotient (i.e. 17+1 = **18**) and 9 is deducted from the remainder (i.e. 11 - 9 = **2**)

*Thus, new quotient is **18** and new remainder is **2**.*

2) 223 divided by 9
Here,
Quotient = *(ab +a)* = (22 + 2) = **24**
Remainder = (2 + 2 + 3) = **7**

3) 321 divided by 9
Here,
Quotient = *(ab +a)* = (32 + 3) = **35**
Remainder = (3 + 2 + 1) = **6**

Transpose Method Division

The method is used, when divisor is a little greater than the base (nearest power of 10).

Transpose = (base – divisor)

Suppose divisor is nearer to base 10, 100, 1000…..etc., then we will use the vedic sutra *"Paravartya Yojayet"*, which means *"transpose and apply"*.

Example-1
Divide *3698* by *112*, and find quotient & remainder?.

As we know that transpose is (base - divisor)
So, we have
Transpose = (100 -112) = $\overline{12}$

1 1 2	3	6	9	8
	↓	3	$\overline{6}$	
$\overline{12}$	$\overline{}$		3 $\overline{6}$	
			$\overline{}$	
	3	3	0	2

Steps-1:
Find out transpose *(100 – 112)* = $\overline{12}$ (bar on top indicates negative value).

Since there are two zero in base, separate two digit of dividend by vertical line for remainder .

Steps-2:

Put the dividend on left column as shown and place transpose below it.

Steps-3:
Now, note down first digit directly as it is, indicated by arrow in second column.

Steps-4:
Then multiply with transpose. $(3 \times -1) = $ **-3** and $(3 \times -2) = $ **-6**

Steps-5:
Write down the multiplication results into second and third column respectively. Then add the column $(6-3) = $ **3**.

Steps-6:
Now, multiply again the transpose from the result of second column i.e. $(3 \times -1) = $ **-3** and $(3 \times -2) = $ **-6** and note down the results into third column (remainder column).

Steps-7:
Add the remainder column $(9 - 6 - 3) = $ **0** and $(8 - 6) = $ **2**.

Thus, we have a **quotient 33** and **remainder 2**.

Crowning Gem Method of Division.

The crowning gem is one-line mental division methods of arithmetic, in this method the actual division is done by single digit. Here, we use vedic sutra *"Dhvaja-anka"* which means (the flag number).

The crowning gem method is generally completed in three basic steps :

- Divide
- Multiply
- Subtract

Example-1
Divide 174 by 53 ?

Let us explain the method step by steps in detail with this example –

Step-1:
Write down the first digit 5 of divisor on the left and remaining digit i.e. 3 on the flag.

	3	17	$_2$4
5			- 9
		3	**15**

Now, we counts the number of digits digit on the flag, in this case there is only one digit, so one right most digit of the dividend (i.e. 4) will be separated by vertical line for remainder position from the remaining digits of dividend for the quotient position (i.e. 17).

Step-2:
Divide 17 by 5,
So, we get 3 as quotient and 2 as remainder

Step-3:
Now, place quotient part below 17 and prefix the remainder 2 just before the 4 on right side in remainder column, then it becomes, 24.

Step-4:
Multiply quotient 3 with flag 3 (i.e. $3 \times 3 = 9$).

Step-5:
Subtract 9 from remainder part (i.e. $24 - 9 = 15$), and write down the remainder 15.
Thus, quotient is 3 and remainder is 15.

Example-2
Divide 235 by 88

	8	23	$_7$5
8			-16
		2	59

Step-1:
Write down the first digit 8 of divisor as base and last digit 8 on flag.

Step-2:
Divide 23 by 8 and place 2 just below 23 and prefix the remainder 7 just before the 5 on right side in remainder column, then it becomes, 75.

Step-3:
Multiply quotient 2 with flag 8 (i.e. 2 × 8 = 16).

Step-4:
Subtract 16 from remainder part (i.e. 75 - 16 = 59).

Step-5:
Check if, remainder is less than the divisor, (yes, here, remainder 59 is less than the divisor 88). So we can declare 59 as final remainder.
Thus, quotient is 2 and remainder is 59.

Practice the following divisions using crowning gem method of division.

i)   Divide 436 by 33
ii)   Divide 432 by 27
iii)   Divide 831 by 56
iv)   Divide 234 by 85
v)   Divide 636 by 38
vi)   Divide 487 by 32
vii)   Divide 336 by 34
viii) Divide 942 by 37
ix)   Divide 436 by 36
x)   Divide 738 by 31
xi)   Divide 593 by 39

Subtraction Method of Division.

In subtraction method you have to subtracts the divisor repeatedly from given number until the remainder is zero, or smaller than the given number (or divisor).

#Rule-1:
Subtracts the divisor repeatedly from given number until the remainder is zero, or smaller than the divisor.

#Rule-2:
Count how many times divisor is subtracted (This will be the quotient)

Example - 1
Find 39 ÷ 6 division using subtraction method.

#Rule-1:
Subtracts the divisor repeatedly from given number until the remainder is zero, or smaller than the divisor.

We start subtracting 6 from 39 until the remainder is zero, or smaller than 6.

$$
\begin{aligned}
39 - 6 &= \mathbf{33} \\
33 - 6 &= \mathbf{27} \\
27 - 6 &= \mathbf{21} \\
21 - 6 &= \mathbf{15} \\
15 - 6 &= \mathbf{9} \\
9 - 6 &= \mathbf{3}
\end{aligned}
$$

#Rule-2:
Count how many times divisor is subtracted (This will be the quotient)

Here, we subtract 6 from 39, six times, and we have 3 as remaining value.

So we have, Quotient = **6** and Remainder = **3**.

Example - 2.
Find $140 \div 35$ division using subtraction method.

We start subtracting 15 from 140 until the remainder is zero, or smaller than 15.

$$140 - 35 = 105$$
$$105 - 35 = 70$$
$$70 - 35 = 35$$
$$35 - 35 = 0$$

Here, we subtract 35 from 140, four times, and we have 0 remaining or remainder value.

So, we have, Quotient = **4** and Remainder = **0**.

Chapter – 8

Tricks for Solving Algebraic Equations

As we have already discussed "Vertically and crosswise" Tricks of multiplication in earlier chapter. In the multiplication of algebraic equations, we follow the same method as in number multiplication using the formula "vertically and crosswise". This ancient technique transcends mere arithmetic, proving to be a powerful tool in simplifying even the most intricate of algebraic expressions. Let us understand with examples.

Algebraic equations with two terms:

Example-1

Solve the algebraic equation $(3x + 2y)(4x + 3y)$.

Step-1:

Write variables x and y and the corresponding coefficients of equations, below them, then use the sutra "Vertical and Crosswise" for 2-digits –

$$(3 \times 4)x^2 + [(3 \times 3)xy + (4 \times 2)xy] + (2 \times 3)y^2$$

Thus, we have $(3x + 2y)(4x + 3y) = 12x^2 + 17xy + 6y^2$

Multiplication of algebraic equations with higher power variable:

Example-1

Solve the algebraic equation $(3x^2 + 2x + 1)(5x^2 + 3x + 2)$

Step-1:

Write variables x^2, x and independent term x^0 and the corresponding coefficients of equations, below them-

$$
\begin{array}{ccc}
x^2 & x^1 & x^0 \\
3 & 2 & 1 \\
5 & 3 & 2
\end{array}
$$

Now use "vertical and crosswise" multiplications for 3-digits, begin from right to left –

i) ii) iii) iv) v)

Steps-2:

i) Vertically on left
  $(3 \times 5) = 15$

ii) Crosswise on left
  $(3 \times 3) + (5 \times 2) = 19$

iii) Crosswise on extreme sides and mid-vertically
  $(3 \times 2) + (5 \times 1) + (2 \times 3) = 17$

iv) Crosswise on right

$(2 \times 2) + (3 \times 1) = 7$

v) Vertically on right

$(1 \times 2) = 2$

Thus, we have

$$
\begin{array}{ccc}
x^2 & x^1 & x^0 \\
3 & 2 & 1 \\
5 & 3 & 2 \\
\hline
\end{array}
$$

15 + 19 + 17 + 7 + 2

Now, add variables to the above coefficients, start from the right most independent terms, next term will be x and then keep increasing the power of variable till the last term-

$15x^4 + 19x^3 + 17x^2 + 7x + 2$

$(3x^2 + 2x + 1)(5x^2 + 3x + 2) = 15x^4 + 19x^3 + 17x^2 + 7x + 2$

Multiplication of an algebraic equation with unequal number of terms:

Example-1

Multiplication of $(3x^2 + x + 2)(x + 3)$

Step-1:

Here x^2, them is missing in second equation, we can put '0' for its coefficient. Then use sutra "vertical and crosswise" multiplications of 3-digits –

Steps-2:

i)   Vertically on left
    $(3 \times 0) = 0$

ii)   Crosswise on left
    $(3 \times 1) + (0 \times 1) = 3$

iii) Crosswise on extreme sides and mid-vertically
    $(3 \times 3) + (0 \times 2) + (1 \times 1) = 10$

iv) Crosswise on right
    $(1 \times 3) + (1 \times 2) = 5$

v) Vertically on right
    $(2 \times 3) = 6$

Thus, we have

$$
\begin{array}{ccc}
x^2 & x^1 & x^0 \\
3 & 1 & 2 \\
0 & 1 & 3 \\
\hline
\end{array}
$$

$$0 + 3 + 10 + 5 + 6$$

Now, add variables to the above coefficients, start from the right most independent terms, next term will be x and then keep increasing the power of variable till the last term-

$$0x^4 + 3x^3 + 10x^2 + 5x + 6$$

$$(3x^2 + x + 2)(x + 3) = 3x^3 + 10x^2 + 5x + 6$$

If the Samuchchaya is Same, it is Equal to Zero

Vedic sutra "Shunyam Samya-Samuchchaya" which means, "*If the Samuchchaya is the Same, it is Equal to Zero*), here the "the term "Samuchhaya" has different meaning in different context of the problem.

Example-1
Solve the equation $5(x - 2) = 3(x - 2)$

#Rule:
If factor $(x - 2)$ is same on both side, therefore it is equated to 0.

i.e.
$(x - 2) = 0$ And $x = 2$

Example-2
Solve $(x + 3)(x + 10) = (x + 5)(x + 6)$

#Rule:
If the product of the independent terms is same in both sides, then common factor is equated to 0.

In given equation $(x + 3)(x + 10) = (x + 5)(x + 6)$ the product of independent terms (non-x terms) are same:
$3 \times 10 \quad = 30$
$5 \times 6 \quad = 30$

Now, since product of the independent terms are the same, therefore, $x = 0$.

Example-3
Solve $1/(3x - 8) + 1/(5x - 8) = 0$

#Rule:
The sum of the denominators of two fractions with same numerical numerator is equated to 0.

$1/(3x - 8) + 1/(5x - 8) = 0$

Thus, the sum of denominators:
$(3x - 8) + (5x - 8) = 0$

So, we have $x = 2$

Example-4
Solve $(2x + 3)/ (5x + 2) = (5x + 2) / (2x + 3)$

#Rule:
If the sum of the numerators and the sum of the denominators in both sides are the same, then that sum is equated to 0.

$(2x + 3)/ (5x + 2) = (5x + 2) / (2x + 3)$

The sum of numerators = sum of the denominators.

Therefore, $(2x + 3) + (5x + 2) = 0$
$7x + 5 = 0$
And $x = -5/7$

If one variable is in ratio with the right-hand side, other one is Zero

Application of vedic sutra - *"Anurupye Shunyamanyat"*, which means (If one variable is in ratio with the right-hand side, other one is Zero). This sutra is applicable in solving a special type of simultaneous simple equations in which the coefficients of 'one' variable are in the same ratio to each other as the independent terms are to each other.

Example-1
Solve the following simultaneous simple equations
$4x + 2y = 16$ and $8x + 4y = 32$

The ratio of x-terms : $4x/8x = \frac{1}{2}$
The ratio of the R.H.S : $16/32 = \frac{1}{2}$

Here, x-terms are in ratio with the right-hand side, the other variable, $y = 0$

Substitute the value of y in above eqn.
We can find-value of x
$4x + 2(0) = 16$
$4x = 16$

Therefore, $x = 2$.

Solving simultaneous simple linear equations, which have the co-efficient of the variables interchanged

#Rule:
By Addition and by Subtraction.

The application of vedic sutra *"Sankalana-vyakalanabhyam"*, which means *(By Addition and by Subtraction)*.

The simple linear equations of type -

$$ax + by = c1 \quad \dots (i)$$
$$bx + ay = c2 \quad \dots (ii)$$

Here, coefficient of x and coefficient of y are identical and vice versa.

Example-1

To solve simultaneous simple linear equations, which have the co-efficient of the variables interchanged.

$$2x - 3y = 4 \quad \dots [1]$$
$$3x - 2y = 6 \quad \dots [2]$$

By addition, we, get
$$(2x - 3y) + (3x - 2y) = 4 + 6$$
$$5x - 5y = 10$$
$$x - y = 2 \quad \dots [3]$$

Now by subtraction", we, get -
$$(2x - 3y) - (3x - 2y) = 4 - 6$$
$$x + y = 2 \quad \dots [4]$$

Now, adding, eqn. **[3]** and **[4]**
We have
$(x - y) + (x + y) = 2 + 2$
$2x = 4$
$x = 2$

Put the value of x in eqn. **[3]**
$x - y = 2$
$y = x - 2$
$y = 2 - 2$
$y = 0$

Practice to solve the following equations

    i) $2x + 3y = 6$
                $4x - y = 5$

    ii) $x + 3y = 3$
                $3x - y = 7$

    iii) $5x + 4y = 20$
                $x - y = 1$

    iv) $6x - 2y = 8$
                $3x + 5y = 7$

    v) $x + 3y = 7$
                $4x + 2y = 14$

To express a fraction as a decimal to all its decimal places.

By vedic sutra - *"Sheshanayanken Charamen"*, which means (*The Remainders by the Last digit*).

Example-1
Express the fraction 1/7 as a decimal to all its decimal places

#Rule:
Add 'zero' to the numerator which makes 1 as 10. If the numerator is less than the denominator, add another 'zero' else proceed to divide, and find out the remainder:

(10)/7	= **1** remainder 3
(30)/7	= **4** remainder 2
(20) /7	= **2** remainder 6
(60) /7	= **8** remainder 4
(40) /7	= **5** remainder 5
(50) /7	= **7** remainder 1

Now note that the remainder is '1' which is same as the numerator '1'. It means we would get the repetition of the answers again and again.

Using the these remainders 3,2,6,4,5 and 1 got above and *multiply* them with the *denominator* '7'

$7 \times 3 = 2\,\mathbf{1}$

$7 \times 2 = 1\,\mathbf{4}$

$7 \times 6 = 4\,\mathbf{2}$

$7 \times 4 = 2\,\mathbf{8}$

$7 \times 5 = 3\,\mathbf{5}$

$7 \times 1 = 0\,\mathbf{7}$

Now, choose the unit digits (right most digits) of the above results of the multiplication as sequence, **142857**

So, you can see the repetition of the decimal after six places.

Therefore, we can write -
    (1/7) = 0.142857 142857 142857 and so on.

Finding the roots of a quadratic equation

Application of Sutra – *"Chalana-kalanabhyam"*, which means (*Sequential motion or "By Calculus"*).

It is used to find the roots of a quadratic equation in calculus and to factorising 3rd, 4th, and 5th degree expressions in differential calculus.

Find the roots of a quadratic equation
$$x^2 - 2x - 4 = 0 \ldots [1]$$

Rule:
By Calculus formula
$$x = \frac{-b \pm \sqrt{b^2 - 4ac}}{2a}$$
Or
$$2ax + b = \pm \sqrt{D} \qquad \ldots [2]$$
where,

$$D = b^2 - 4ac$$

Now, put the value of equation [1] in equation [2]

We, get

$$2x - 2 = \pm \sqrt{20}$$

$$2x - 2 = \pm 2\sqrt{5}$$

$$2(x - 1) = \pm 2\sqrt{5}$$

$$(x - 1) = \pm \sqrt{5}$$

Thus, the roots of a quadratic [1] equation are

$$x = + \sqrt{5} + 1 \quad \text{and} \quad x = - \sqrt{5} + 1$$

Example-1

Find the roots of a quadratic equation

$$x^3 - 3x + 1 = 0 \ \dots [1]$$

By Calculus formula

$$2ax + b = \pm \sqrt{b^2 - 4ac}$$

$$2x - 3 = \pm \sqrt{5}$$

$$2x = 3 \pm \sqrt{5}$$

Thus, the roots of a quadratic [1] equation are

$$x = \frac{3 \pm \sqrt{5}}{2}$$

The factorization of quadratic equations

By vedic sutra *"Vyasti-samashti"*, which means (*Whole as One and One as Whole*).

Example-1: Factorizing the following quadratic equation

$$x^2 + 8x + 15 = 0$$

#Rules:

i) Make factors of independent number 15 such that its sum becomes equal to the middle independent number 8.

ii) We can write factor of 15 as (3 × 5), here you can see that sum of factor (3 + 5) is equal to 8.

iii) Thus we can write the above quadratic equation as -

$$(x + 3)(x + 5)$$

From the above example, we can conclude the factorization of quadratic equation and writing a general equation. Suppose, we have quadratic equation

$$x^2 + bx + c = 0$$

Then factor of independent term C = p × q, such that,

$$B = (p + q)$$

Then quadratic equation can be written as -

$$x^2 + x(p + q) + pq = 0$$

Therefore, factors of above quadratic equation can be written as - (x + p)(x + q)

Chapter – 9

Square-Roots of Perfect Squares

Square-root of a number is a value that, when multiplied by itself, gives the original number in general method. For example, the square root of 9 is 3 because 3 multiplied by itself (3×3) equals 9.

Perfect numbers are positive integers that are equal to the sum of their proper divisors (excluding the number itself). The square root of a perfect number is a rational number if the perfect number is a perfect square. For example, 4 is a perfect square, and its square root is 2. However, not all perfect numbers are perfect squares, so their square roots may be irrational. The study of perfect numbers involves understanding their properties and relationships with factors and divisors.

Here, we can find out the square root of perfect number by using vedic tricks. You must remember the following facts about perfect numbers:

#Rule-1:
Perfect numbers ends in 0, 1, 4, 5, 6 and 9.

#Rule-2:
For even numbers, last two digits must be divisible by 4.

#Rule-3:

If number ends in 6 then second last digit must be odd.

To find out the square-roots of perfect numbers in vedic mathematics, we use the sutra *"Vilokanam"* which means *'By mere Observation'*.

Let us write the squares of the first 10 natural numbers:

Number	Square	Last digit of Square
1^2	1	1
2^2	4	4
3^2	9	9
4^2	16	6
5^2	25	5
6^2	36	6
7^2	49	9
8^2	64	4
9^2	81	1
10^2	100	00

We conclude or made observations from the above table:

- If the ends digit of the square is 1, then the last digit of its square root will be: **1** or **9**.
- If the ends digit of the square is 4, then the last digit of its square root will be: **2** or **8**.

- If the ends digit of the square is 5, then the last digit of its square root will be: **5**.
- If the ends digit of the square is 6, then the last digit of its square root will be: **4** or **6**.
- If the ends digit of the square is 9, then the last digit of its square root will be: **3** or **7**.

You can prepare a table for your convenience form the above observations

Observation table

If last digit of Square is	1	4	9	6	5	0
Square root will be	1, 9	2, 8	3, 7	4, 6	5	0

After observing the above table the finding of square root of perfect number becomes much easier. Remember, the number of digits in the answer of the square root, will be equal to the total number of groups.

#Rule-1:
Make number partitions of two digits from right to left.

#Rule-2:
We, can find out the possible last digit of square root, simply by looking at observation table.

#Rule-3:

Find out the first digit of the answer by observing the square of the number in the next group on the left.

If the number in group lies between the squares of two numbers, then take smaller number. Now there are two possible answer of square root.

#Rule-4:

If the multiple of the chosen smaller number and it's next number is less than the number in a group, then we take smaller possible answer, otherwise we take greater possible number as final answer.

Example-1

Find the square root of $\sqrt{(46\ 24)}$.

Steps to find out the square root of 4-digit number

Step-1:

Make number partitions of two digits from right to left.

$\sqrt{(\overline{46}\ \overline{24})}$

Step-2:

The number $\overline{24}$ (second group), has it's end digit 4. So the possible last digits of the square root is either **2** or **8** (*look at the observation table*).

Step-3:

Now to find out the first digit of the square root, from the observation table. Here, you can see the square of $\overline{46}$, lies between square of ($6^2 = 36$ & $7^2 = 49$). Therefore, the possible square root digits are 6 & 7, where, we shall always choose the smaller digit.

Now, there are two possibilities of the final answer either **62** or **68** (from *step-2* and *step-3*)

Step-4:
Since, the multiple ($6 \times 7 = 42$) is less than the given number $\overline{46}$. So, we have to take bigger number 68 from the *step-3*.

Thus, the square root of $\sqrt{(46\ 24)} = $ **68 Ans.**

Practice the Following Square-root of Perfect Numbers

1) $\sqrt{(44\ 89)}$ 2) $\sqrt{(40\ 96)}$

3) $\sqrt{(12\ 96)}$ 4) $\sqrt{(67\ 24)}$

5) $\sqrt{(27\ 04)}$ 6) $\sqrt{(20\ 25)}$

7) $\sqrt{(39\ 69)}$ 8) $\sqrt{(54\ 76)}$

9) $\sqrt{(94\ 09)}$ 10) $\sqrt{(7921)}$

11) $\sqrt{(5929)}$ 12) $\sqrt{(4356)}$

Find the Square Roots "By Repeated Subtraction of Odd Numbers".

We know from the sum property of odd numbers, that the sum of the first *n* odd numbers is:

$$1 + 3 + 5 + 7 + \ldots\ldots\ldots + (2n{-}1) = n^2.$$

We can see from the above expression, that the square number is the sum of successive odd numbers starting from 1. Therefore, you can find square root of a number by repeatedly subtracting successive odd numbers from the given number, until *n* becomes zero.

Example -1.

Find square root of 121 by subtraction method.

Simple mental steps

Take number to find out the square root

$n = 121$

Subtracting odd numbers starting from 1, until the number becomes 0.

1. $121 - 1 = 120$
2. $120 - 3 = 117$
3. $117 - 5 = 112$
4. $112 - 7 = 105$
5. $105 - 9 = 96$
6. $96 - 11 = 85$
7. $85 - 13 = 72$

8. $72 - 15 = 57$
9. $57 - 17 = 40$
10. $40 - 19 = 21$
11. $21 - 21 = 0$

Since, 11 odd numbers were used to subtract, hence the square root of 121 is **11**.

Example -2

Find square root of 25 by subtraction method.
Take number to find out the square root

$n = 25$

Subtracting odd numbers starting from 1, until the number becomes 0.

1) $25 - 1 = 24$
2) $24 - 3 = 21$
3) $21 - 5 = 16$
4) $16 - 7 = 9$
5) $9 - 9 = 0$

Here, we used, 5 odd numbers to subtract the number so that it become zero, hence the square root of 25 is **5**.

Chapter – 10

Cube-Roots of Perfect Cubes

Cube-root of a number is a value that, when multiplied by itself three times, gives the original number in general method. The cube roots follow the same pattern as square roots. To find out cube-roots of perfect numbers in vedic mathematics, we follow the same methods and procedures as we have apply in square root of perfect number, we use the same sutra *"Vilokanam"*, which means (*By mere Observation*).

Let us write the cubes of the first 9 natural numbers:

Number	Cubes	Last digit of Cubes
1^3	1	1
2^3	8	8
3^3	27	7
4^3	64	4
5^3	125	5
6^3	216	6
7^3	343	3
8^3	512	2
9^3	729	9

Observation table derived from the cube of 9-natural numbers:

If a Cubes of Number End in	Then Cubes Roots will End in
1	1
2	8
3	7
4	4
5	5
6	6
7	3
8	2
9	9

From the above table the finding of cube roots of perfect number becomes much easier. To find out the cube roots of a number we make group of three digits starting from the right. The number of digits in the answer of cube roots will be equal to the number of groups.

Now, apply vedic sutra "Vilokanam" which means (*By mere Observation*), from the above observation table, we can find out the possible last digit of cube root, simply by looking at the last digit of the cube of a number and we can find out the first digit of cube root by observing the second group of number on the left.

Example: Find the cube root of $\sqrt[3]{}$ (24389).

Steps to find out the cube root of 5-digit number (24389).
1) Make number partitions of three digits start from right
$$\sqrt[3]{(\overline{24}\ \overline{389})}$$

Now, apply the sutra "Vilokanam" which means (*By mere Observation*).

2) We have two groups of number ($\overline{24}\ \overline{389}$), in second group $\overline{389}$ the last digit of number is 9, so the second digit of cube root will be 9.

3) Now we observe the first groups of number $\overline{24}$ whose cube lies between ($2^3 = 8$ & $3^3 = 27$). We can see that 27 is greater than 24, so we can take smaller number as answer of first group (i.e., 2).

$$\sqrt[3]{}\ (24389) = 29\ \textbf{Ans.}$$

Practice the Following Cube-root of Perfect Numbers
1) $\sqrt[3]{(15625)}$ 2) $\sqrt[3]{(19683)}$

3) $\sqrt[3]{(32768)}$ 4) $\sqrt[3]{(46656)}$

5) $\sqrt[3]{(74088)}$ 6) $\sqrt[3]{(97336)}$

7) $\sqrt[3]{(29791)}$ 8) $\sqrt[3]{(64000)}$

9) $\sqrt[3]{(27000)}$ 10) $\sqrt[3]{(21952)}$

Chapter – 11

Vedic Square and Numbers Patterns

The multiplication table for 9 × 9 matrix has many interesting facts and properties, it is created by multiplying value of row and corresponding column.

0	1	2	3	4	5	6	7	8	9
1	1	2	3	4	5	6	7	8	9
2	2	4	6	8	10	12	14	16	18
3	3	6	9	12	15	18	21	24	27
4	4	8	12	16	20	24	28	32	36
5	5	10	15	20	25	30	35	40	45
6	6	12	18	24	30	36	42	48	54
7	7	14	21	28	35	42	49	56	63
8	8	16	24	32	40	48	56	64	72
9	9	18	27	36	45	54	63	72	81

From the multiplication table we can generate a vedic square by replacing every number in the cell by its digit sum. In other word a vedic square is a typical 9×9 square matrix, where the entry in each cell is the digital root of the product of the row and column headings.

1	2	3	4	5	6	7	8	9
2	4	6	8	1	3	5	7	9
3	6	9	3	6	9	3	6	9
4	8	3	7	2	6	1	5	9
5	1	6	2	7	3	8	4	9
6	3	9	6	3	9	6	3	9
7	5	3	1	8	6	4	2	9
8	7	6	5	4	3	2	1	9
9	9	9	9	9	9	9	9	9

The vedic square is a tool that simplifies calculations by utilizing patterns and relationships between numbers within the grid.

Vedic Square has been coded with unique colours for each number.

1	2	3	4	5	6	7	8	9
2	4	6	8	1	3	5	7	9
3	6	9	3	6	9	3	6	9
4	8	3	7	2	6	1	5	9
5	1	6	2	7	3	8	4	9
6	3	9	6	3	9	6	3	9
7	5	3	1	8	6	4	2	9
8	7	6	5	4	3	2	1	9
9	9	9	9	9	9	9	9	9

Reflection patterns generated about vertical and horizontal axis respectively for the above Vedic Square.

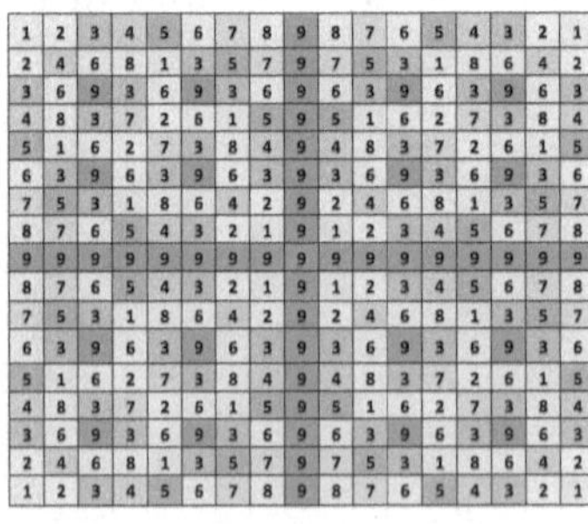

We Can Generate 9-Unique Patterns from Vedic Square for Individual Numbers

All 1s

1	2	3	4	5	6	7	8	9
2	4	6	8	1	3	5	7	9
3	6	9	3	6	9	3	6	9
4	8	3	7	2	6	1	5	9
5	1	6	2	7	3	8	4	9
6	3	9	6	3	9	6	3	9
7	5	3	1	8	6	4	2	9
8	7	6	5	4	3	2	1	9
9	9	9	9	9	9	9	9	9

All 2s

1	2	3	4	5	6	7	8	9
2	4	6	8	1	3	5	7	9
3	6	9	3	6	9	3	6	9
4	8	3	7	2	6	1	5	9
5	1	6	2	7	3	8	4	9
6	3	9	6	3	9	6	3	9
7	5	3	1	8	6	4	2	9
8	7	6	5	4	3	2	1	9
9	9	9	9	9	9	9	9	9

All 3s

1	2	3	4	5	6	7	8	9
2	4	6	8	1	3	5	7	9
3	6	9	3	6	9	3	6	9
4	8	3	7	2	6	1	5	9
5	1	6	2	7	3	8	4	9
6	3	9	6	3	9	6	3	9
7	5	3	1	8	6	4	2	9
8	7	6	5	4	3	2	1	9
9	9	9	9	9	9	9	9	9

All 4s

1	2	3	4	5	6	7	8	9
2	4	6	8	1	3	5	7	9
3	6	9	3	6	9	3	6	9
4	8	3	7	2	6	1	5	9
5	1	6	2	7	3	8	4	9
6	3	9	6	3	9	6	3	9
7	5	3	1	8	6	4	2	9
8	7	6	5	4	3	2	1	9
9	9	9	9	9	9	9	9	9

All 5s

1	2	3	4	5	6	7	8	9
2	4	6	8	1	3	5	7	9
3	6	9	3	6	9	3	6	9
4	8	3	7	2	6	1	5	9
5	1	6	2	7	3	8	4	9
6	3	9	6	3	9	6	3	9
7	5	3	1	8	6	4	2	9
8	7	6	5	4	3	2	1	9
9	9	9	9	9	9	9	9	9

All 6s

1	2	3	4	5	6	7	8	9
2	4	6	8	1	3	5	7	9
3	6	9	3	6	9	3	6	9
4	8	3	7	2	6	1	5	9
5	1	6	2	7	3	8	4	9
6	3	9	6	3	9	6	3	9
7	5	3	1	8	6	4	2	9
8	7	6	5	4	3	2	1	9
9	9	9	9	9	9	9	9	9

All 7s

1	2	3	4	5	6	7	8	9
2	4	6	8	1	3	5	7	9
3	6	9	3	6	9	3	6	9
4	8	3	7	2	6	1	5	9
5	1	6	2	7	3	8	4	9
6	3	9	6	3	9	6	3	9
7	5	3	1	8	6	4	2	9
8	7	6	5	4	3	2	1	9
9	9	9	9	9	9	9	9	9

All 8s

1	2	3	4	5	6	7	8	9
2	4	6	8	1	3	5	7	9
3	6	9	3	6	9	3	6	9
4	8	3	7	2	6	1	5	9
5	1	6	2	7	3	8	4	9
6	3	9	6	3	9	6	3	9
7	5	3	1	8	6	4	2	9
8	7	6	5	4	3	2	1	9
9	9	9	9	9	9	9	9	9

All 9s

1	2	3	4	5	6	7	8	9
2	4	6	8	1	3	5	7	9
3	6	9	3	6	9	3	6	9
4	8	3	7	2	6	1	5	9
5	1	6	2	7	3	8	4	9
6	3	9	6	3	9	6	3	9
7	5	3	1	8	6	4	2	9
8	7	6	5	4	3	2	1	9
9	9	9	9	9	9	9	9	9

If we draw a line and connect each of these numbers, we will have different shape for each square.

The numbers patterns generated from the digital roots below and above the value of number 5, we get a beautiful pattern.

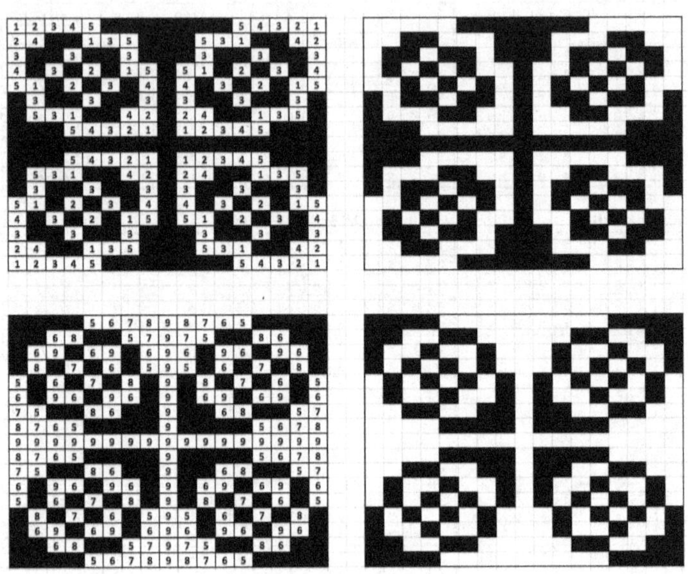

The numbers pattern generated from odd and even value of digital roots in vedic square.

By assigning black colour to odd numbers and keeping blank cells for all even numbers and vice-versa, you can see beautiful patterns looks like, QR code used in digital transactions.

However, this not an actual QR-code, an actual QR codes, are basically generated by specific data or details to be stored in it, which is converted into binary format, where each character or information is represented by a series of 0s and 1s.

These, binary data along with error codes, is structured into a specific format. The code generator performs a series of mathematical calculations and computer algorithms to create the pattern of black and white dots that form the QR codes. Those black and white dots are arranged in square grid patterns.

QR codes come in different versions and patterns, which vary in size and amount of data they can hold. The masking techniques are used, which alters the appearance of QR codes. The black area make QR scanner easy to identify the code boundaries.

Chapter – 12

Magic Squares

A magic square is a square grid of numbers where the sum of each row, each column, and its main diagonals are the same. The sum of the horizontal, vertical, and diagonal lines is called the magic sum or magic constant. It is basically a mathematical and recreational puzzle often used for entertainment.

Now, suppose, square is created for $n \times n$ grid of numbers, where n is an odd number.

$$\text{Magic Sum (MS)} = \frac{n(n^2 + 1)}{2}$$

With this formula you can make magic squares of different orders. If we subtract each number from $(n^2 + 1)$, we get another magic square, and this is called the complementary magic square.

Steps to Create a Magic Square:

Placement Rules:
- Begin by placing the number 1 in the middle of the top row.
- Move diagonally up and to the right to place the next number.

- If moving outside the grid, wrap around to the opposite side (toroidal movement).
- If the cell is already occupied, place the next number directly below the last placed number.

Fill the Grid:

Continue placing numbers sequentially $(2, 3, 4, ..., n^2)$ according to the rules until the grid is completely filled.

Verification:

Verify that the sums of numbers in each row, each column, and both main diagonals equal the same constant value (known as the magic constant).

Example of Creating (3×3) Magic Square:

8	1	6
3	5	7
4	9	2

Steps for creating 3×3 magic square:
1. Start with 1 in the middle of the top row.
2. Move diagonally up and to the right to place subsequent numbers.
3. If a move would take you outside the grid, wrap around to the opposite side.

4. If a cell is already occupied, place the next number directly below the last placed number.
5. Continue until the grid is filled with numbers from 1 to 9.

The calculation of magic sum or magic constant for 3×3 square is given by –

$$MS = \frac{3(3^2+1)}{2} = 15$$

Now, we can verify the above magic sum for each row, column and main diagonal.

Row-1: $8 + 1 + 6 = 15$
Row-2: $3 + 5 + 7 = 15$
Row-3: $4 + 9 + 2 = 15$

Column-1: $8 + 3 + 4 = 15$
Column-2: $1 + 5 + 9 = 15$
Column-3: $6 + 7 + 2 = 15$

Diagonal-1: $8 + 5 + 2 = 15$
Diagonal-2: $6 + 5 + 4 = 15$

Creating a Complementary Magic Square

#Rule-1: If we subtract each number from $(n^2 + 1)$ in the normal magic square, we get another magic square, and this is called the complementary magic square.

The original Magic Square

8	1	6
3	5	7
4	9	2

From #Rule-1: Subtract each number from $(n^2 + 1)$
Here, $n = 3$
Therefore, $(n^2 + 1) = (9+1) = 10$

So, subtract **10** from each number in normal magic square, to form the complementary magic square

2	9	4
7	5	3
6	1	8

Chapter – 13

Special Section for Competitive Exams

Number Theory

Number theory is a branch of mathematics that deals with the properties and relationships of numbers, particularly integers. It explores concepts such as prime numbers, divisibility, modular arithmetic, and Diophantine equations. Number theory has applications in cryptography, computer science, and various other fields.

It is generally, the study of the set of positive whole numbers 1, 2, 3,4,5,….. which are often called the set of natural numbers. It is a branch of pure mathematics devoted primarily to the study of the integers and arithmetic functions. Sometimes called *higher arithmetic*.

In number theory, you study the relationships between different types of numbers and various properties of the integers. The natural numbers are classified into various types.

Odd Numbers:

Positive integers that cannot be divided into two equal parts (e.g. 1, 3, 5, 7, 9, 11, . . .etc.).

Even Numbers:

Positive integers that can be divided into two equal parts (e.g. 2, 4, 6, 8, 10, ...etc.)

Square Numbers:

Number is obtained when an integer is multiplied by itself. (e.g. 1, 4, 9, 16, ...etc.)

Cube Numbers:

Number is obtained when an integer is multiplied by itself three times.
(e.g. 1, 8, 27, 64, ...etc.)

Prime Numbers:

Number that can be divided exactly only by itself.
(e.g. 2, 3, 5, 7, 11, ...etc.)

Euclid's division lemma

Euclid's division lemma is a fundamental concept in number theory. It is also commonly referred to as the Division Algorithm.

It essentially states that for any two positive integers 'a' and 'b', where 'b' is not equal to zero, there exist unique integers 'q' (quotient) and 'r' (remainder) such that '$a = bq + r$' and '$0 \leq r < |b|$'. This lemma forms the basis for many important concepts in number theory, including divisibility, greatest common divisor (GCD), and modular arithmetic.

Checking Divisibility By Numbers

The divisibility tests are fundamental tools in vedic mathematics used to quickly determine if one number can be evenly divided by another. These tests provide efficient shortcuts to avoid the need for traditional long division, especially when dealing with large numbers.

By mastering these tests, mathematicians and students alike can streamline their calculations, saving time and effort. In this chapter, we will explore various divisibility tests, ranging from basic rules for divisibility by small numbers to more advanced techniques for larger ones. By understanding and applying these tests, readers will gain valuable insights into the underlying patterns and properties of numbers.

Checking Divisibility By 2 to 12

Divisibility by 2:
A number is divisible by 2 if its last digit is even (0, 2, 4, 6, or 8).

Example-1
Check if 819748 is divisible by 2.

Since, a number is divisible by 2 if its last digit is even number or 0.

The last digit of given number is 8, Therefore, it is divisible by 2.

Example-2
Check if 2453076 is divisible by 2.

Since, a number is divisible by 2 if its last digit is even number or 0.

The last digit of given number is 6, Therefore, it is divisible by 2.

Divisibility by 3:
A number is divisible by 3 if the sum of its digits is divisible by 3.

Example-1
Check if 819 is divisible by 3 using digit sum.

Digit sum of 819: 8 + 1 + 9 = 18
And digit sum of 18: 1 + 8 = 9
Since 9 is divisible by 3

So, 819 is also divisible by 3.

Example-2
Check if 12345 is divisible by 3 using digit sum.

Digit sum of 12345: 1 + 2 + 3 + 4 + 5 = 15
Since 15 is divisible by 3

So, 12345 is also divisible by 3.

Divisibility by 4:
A number is divisible by 4 if the number formed by its last two digits is divisible by 4.

The divisibility by number 4 can also be check by application of Vedic Sutra *"The Ultimate and Twice the Penultimate"*. If the result is divisible by 4 then the given number will also be divisible by 4. Otherwise it is not divisible by 4.

Example-1
Check if the number 348 is divisible by 4.

Here we can have from the formula *"The Ultimate and Twice the Penultimate"*
8 plus twice 4, which is 16.

Since the result is divisible by 4, therefore, the given number 348 is also divisible by 4.

Example-2
Check if 1296 is divisible by 4.

Here we can have from the formula *"The Ultimate and Twice the Penultimate"*
6 plus twice 9, which is 24.

Since the result is divisible by 4, therefore, the given number 1296 is also divisible by 4.

Example-3
Check if 5684 is divisible by 4.

Here we can have from the formula *"The Ultimate and Twice the Penultimate"*

4 plus twice 8, which is 20.

Since the result is divisible by 4, therefore, the given number 5684 is also divisible by 4.

Example-4:
Check if 792 is divisible by 4.

Here we can have from the formula *"The Ultimate and Twice the Penultimate"*

2 plus twice 9, which is 20.

Since the result is divisible by 4, therefore, the given number 792 is also divisible by 4.

Example-5:
Check if 79**23** is divisible by 4.

Here we can have from the formula *"The Ultimate ·and Twice the Penultimate"*

3 plus twice of the number 2, which is 7.

Since the result is not divisible by 4, therefore, the given number 79**23** is **not** divisible by 4.

Divisibility by 5:

A number is divisible by 5 if its last digit is either 0 or 5.

Example-1:
Check if 7925 is divisible by 5.

Since, last digit of given number is 5
Therefore, 7925 is divisible by 5.

Example-2:
Check if 73547290 is divisible by 5.

Since, last digit of given number is 0
Therefore, 73547290 is divisible by 5.

Divisibility by 6:

A number is divisible by 6 if it is divisible by both 2 and 3.

Example-1:
Check if 127926 is divisible by 6.

Since, last digit of given number is even number i.e. 6
Therefore, 127926 is **divisible by 2**.

The digit sum of $127926 = 1 + 2 + 7 + 9 + 2 + 6 = 27$
And the digit sum of $27 = 2 + 7 = 9$
Therefore, 127926 is also **divisible by 3**.

Thus, the 127926 is **also divisible by 6.**

Divisibility by 7:

For larger numbers, there's a rule called "The doubling and subtracting method." Double the last digit and subtract it from the remaining number. If the result is divisible by 7, then the original number is also divisible by 7.

Example-1:
Check if 399 is divisible by 7.

Double the last digit and subtract it from the remaining number.
$2 \times 9 = 18$
The remaining part of given number is 39, now subtract 18 from it.

$$39 - 18 = 21$$

Since the result 21 is divisible by 7, therefore, the original number 399 is also divisible by 7.

Divisibility by 8:

A number is divisible by 8 if the number formed by its last three digits is divisible by 8.

Example-1:
Check if 645104 is divisible by 8.

The last three digit of the given number is 104, which is divisible by 8.

Therefore, the original number 645104 is also divisible by 8.

Divisibility by 9:

A number is divisible by 9 if the sum of its digits is divisible by 9.

Example-1
Check if 6543 is divisible by 9 using digit sum.

Digit sum of 6543: $6 + 5 + 4 + 3 = 18$
Since 18 is divisible by 9,
Therefore, 6543 is also divisible by 9.

Divisibility by 10:

A number is divisible by 10 if its last digit is 0.

Example-1:
Check if 397649240 is divisible by 10.

The last digit of the given number is 0.

Therefore, the original number 397649240 is also divisible by 10.

Divisibility by 11:

The difference between the sum of digits at even places and the sum of digits at odd places of a number should be divisible by 11.

If the result end up with 0 or 11 or multiple of 11 then the number is divisible by 11. Otherwise the number is not divisible by 11.

In vedic mathematics we can apply simple sutra *"By Addition and by Subtraction"*.

Example-1
Check if the number 3803382 is divisible by 11.

Add all the digits in the odd positions, starting with the units, and all the digits in the even positions and subtract the smaller result from the larger result.

Sum of all the digits in odd positions: $(2 + 3 + 0 + 3) = 8$
Sum of all the digits in the even positions: $(8 + 3 + 8) = 19$

By subtracting the smaller result from the larger result, we got $19 - 8 = 11$

Since the result end up with 11, so the number 3803382 is divisible by 11.

Example-2
Check if the number 1234 is divisible by 11.

Sum of all the digits in odd positions: $(4 + 2) = 6$
Sum of all the digits in the even positions: $(3 + 1) = 4$

By subtracting the smaller result from the larger result, we got $6 - 4 = 2$

Since the result end up with number 2, which is not divisible by 11, so the number 1234 is not divisible by 11.

Divisibility by 12:

A number is divisible by 12 if it is divisible by both 3 and 4.These methods allow for quick mental calculations and divisibility checks without having to resort to long division.

Example-1:

Check if 979248 is divisible by 12.

Check whether it is divisible by 3

The digits sums $= 9 + 7 + 9 + 2 + 4 + 8 = 39$
And the digits sums of $39 = 3 + 9 = 12$

Since, 12 is divisible by 3, therefore the given number 979248 is also divisible by 3.

Now, check divisibility by number 4

The last digit of number is 48
Since the last two digits is divisible by 4, therefore the given number 979248 is also divisible by 4.

Now because, number 979248 is divisible by both 3 and 4, therefore it is also divisible by 12.

Digit Sums

The digit sum of a number is the sum of its individual digits. For example, the digit sum of 123 would be $1 + 2 + 3 = 6$.

It's a way to analyse the properties of numbers and can be useful in various mathematical calculations and puzzles.

If the sum of the digits is greater than 9, then sum the digits of the result again until the result is less than 10. Here, the numbers 1, 2, 3, 4, 5, 6, 7, 8, 9, 0 are all digits. Bigger numbers can be reduced to the single digit by adding it's constituents digits. Understanding the digits sum of the of numbers serves as a foundational tool in determining divisibility.

Example-1: The addition of nine natural numbers is equal to 45, find the smallest number with distinct digits, whose digits sum add up to 45?

Select the correct answer -
   i) 123456789   ii) 132546789
   iii) 123555789   iv) 555555555

Solution:

Let's find the digit sum of each number

    i) 1+2+3+4+5+6+7+8+9 = **45**

    ii) 1+3+2+5+4+6+7+8+9 = **45**

    iii) 1+2+3+5+5+5+7+8+9 = **45**

    iv) 5+5+5+5+5+5+5+5+5 = **45**

We can see in (i) and (ii) all the digits in this number are distinct, and sum is equal to 45.

Whereas (iii) and (iv) does not satisfy the requirement of distinct digits, because digit 5 is repeating.

Now, we can easily find out the smallest number i.e. 123456789 with distinct digits and whose digits sum add up to 45.

Thus, the correct answer is (i) 123456789.

Amazing Short Tricks of Simplifying Various Mathematical Puzzles, in Competitive Exams

In competitive exams, various short tricks could be game-changer. These short tricks simplifies the calculation process, and save our precious time and enhance efficiency during examinations.

By understanding and employing such short tricks, we can solve big mathematical puzzles with greater ease and confidence, ultimately improving their performance in competitive exams.

Mathematical equations frequently asked in competitive exams

If $x^n + \dfrac{1}{x^n} = k$, then find the value of $x^n - \dfrac{1}{x^n} = ?$

#Identity-1:

If $\quad x^n + \dfrac{1}{x^n} = k$

Then $x^n - \dfrac{1}{x^n} = \sqrt{(k^2 - 4)}$

#Identity-2:

If $\quad x^n - \dfrac{1}{x^n} = k$

Then $x^n + \dfrac{1}{x^n} = \sqrt{(k^2 + 4)}$

#Identity-3:

For (n =1)
$$x + \frac{1}{x} = k$$
Then $x - \frac{1}{x} = \sqrt{(k^2 - 4)}$

#Identity-4:

If $x - \frac{1}{x} = k$
Then $x + \frac{1}{x} = \sqrt{(k^2 + 4)}$

Example-1 If $(x^8 + \frac{1}{x^8}) = 11$, (where $x > 0$), then find the value of $\left(x^8 - \frac{1}{x^8}\right) = ?$

The correct answer is -

    i)   2

    ii) $\sqrt{(11)}$

    iii) $\sqrt{(17)}$

    iv) None of the above

Solution:

From #Identity-1:

    If $x^n + \frac{1}{x^n} = k$

    Then $x^n - \frac{1}{x^n} = \sqrt{(k^2 - 4)}$

Now,

If $x^8 + \frac{1}{x^8} = 11$,

Then $x^8 - \frac{1}{x^8} = \sqrt{(11^2 - 4)} = \sqrt{(121 - 4)} = \sqrt{(17)}$

So, we could write $(x^8 - \frac{1}{x^8}) = \sqrt{(17)}$ **Ans.**

Example-2 If $(x^{256} + \frac{1}{x^{256}}) = 95$, (where $x > 0$),
then find the value of $(x^{256} - \frac{1}{x^{256}}) = ?$

The correct answer is -

   v)   256
   vi) $\sqrt{(9021)}$
   vii) $\sqrt{(9025)}$
   viii) None of the above

Solution:

From #Identity-1:

If $x^n + \frac{1}{x^n} = k$

Then $x^n - \frac{1}{x^n} = \sqrt{(k^2 - 4)}$

So, we can write

$(x^{256} - \frac{1}{x^{256}}) = \sqrt{(95^2 - 4)}$ ....(i)

Now to find out the square of 95^2 we, apply, Vedic tricks of squaring numbers that ending in 5.

Multiply first digit by Ekadhikena Purvena "One more than the previous"

So, we can write $9 \times 10 = 90$ (first part of answer)
Then place 25 in the end

So, we have $95^2 = 9025$
Now, from eqn. (i)

$$\left(x^{256} - \frac{1}{x^{256}}\right) = \sqrt{(9025 - 4)}$$

$$\left(x^{256} - \frac{1}{x^{256}}\right) = \sqrt{(9021)}$$

Addition Tricks for the Series of 10 - Continuous Natural Numbers.

#Rule-1: Add 4 to the first number (you will get the first part of the answer)

#Rule-2: Just place number 5 in end to complete the answer.

Examples-1:
$$1 + 2 + 3 + 4 + 5 + \ldots\ldots + 9 + 10$$

The correct answer is -

    i)   5
    ii)   55
    iii)   555
    iv)   None of the above

Step-1:
Add 4 to the first number
$1 + 4 = \mathbf{5}$

Step-2:
Place 5 at the end
From *Step-1 & Step-2*, we get **55 Ans.**

Examples-2:
$$211 + 212 + 213 + 214 + 215 + \ldots\ldots + 219 + 220$$

The correct answer is -

    i)   1555
    ii)   2055

iii)   2155

iv)   None of the above

Step-1:
Add 4 to the first number
$211 + 4 = \textbf{215}$

Step-2:
Place 5 at the end

From *Step-1 & Step-2,* we get **2155 Ans.**

Examples-3:
$3061 + 3062 + 3063 + 3064 + 3065 + \ldots + 3069 + 3070$

The correct answer is -

i)   30555

ii)   30655

iii)   30755

iv)   None of the above

Step-1:
Add 4 to the first number
$3061 + 4 = \textbf{3065}$

Step-2:
Place 5 at the end

From *Step-1 & Step-2,* we get **30655 Ans.**

Examples-4:
 3076 + 3077 + 3078 + 3079 + 3080 + ... + 3084 + 3085

The correct answer is -
 i)   30605
 ii)   30705
 iii) 30805
 iv)   None of the above

Step-1:
Add 4 to the first number
$(3076 + 4) = \mathbf{3080}$

Step-2:
Place 5 at the end of 3080

So, we get **30805 Ans.**

Examples-5:
 3581 + 3582 + 3583 + 3584 + 3585 + ... + 3589 + 3590

The correct answer is -
 i)   30855   ii) 35855
 iii) 40855   iv) None of the above

Steps:
Add 4 to the first number i.e. $(3581 + 4) = \mathbf{3585}$
Now, place 5 at the end 3585

So, we get the Answer **35855.**

Amazing Tricks of Addition and Subtraction of Fraction Series.

Addition Tricks

Examples-1: Find the sum of given fraction series

$$\frac{1}{3} + \frac{1}{6} + \frac{1}{12} + \frac{1}{20} + \frac{1}{30} + \frac{1}{42} + \frac{1}{56} = ?$$

Select the correct answer -

(i) $\frac{17}{12}$ (ii) $\frac{17}{24}$ (iii) $\frac{17}{48}$ (iv) None of the above

Solution:

We can re-write the series as -

$$\frac{1}{3} + \frac{1}{2\times3} + \frac{1}{3\times4} + \frac{1}{4\times5} + \frac{1}{5\times6} + \frac{1}{6\times7} + \frac{1}{7\times8} \qquad \text{........ (i)}$$

As we know that

$$\frac{1}{2 \times 3} = \frac{1}{2} - \frac{1}{3}$$

In more generalize way we can write as

$$\frac{1}{p \times q} = \frac{1}{p} - \frac{1}{q}$$

where, $(p + 1) = q$ (an integer)

Thus, can easily write eqn. (i) as

$$\frac{1}{3} + \left(\frac{1}{2} - \frac{1}{3}\right) + \left(\frac{1}{3} - \frac{1}{4}\right) + \left(\frac{1}{4} - \frac{1}{5}\right) + \left(\frac{1}{5} - \frac{1}{6}\right) +$$

$$\left(\frac{1}{6} - \frac{1}{7}\right) + \left(\frac{1}{7} - \frac{1}{8}\right) \qquad \text{........ (ii)}$$

We, know that, the similar plus and minus terms will cancelled out, and the eqn.(ii), reduced to -

$$\frac{1}{3} + \frac{1}{2} - \frac{1}{8}$$ .... (iii)

The least common multiple (LCM) of 3, 2, and 8 is 24

Now, we can add and subtract the fraction in eqn.(iii)

$$\frac{8}{24} + \frac{12}{24} - \frac{3}{24} = \frac{8+12-3}{24} = \frac{17}{24}$$

Thus, the sum of given fraction series is $\frac{17}{24}$ **Ans.**

Amazing Tricks of Simplifying Surds

In mathematics, a surd is an expression containing a square root, cube root, or any other root that cannot be simplified to remove the root symbol. In other words, a surd is an irrational number that can be written in root form.

Surds often appear in algebra and geometry and are useful in various mathematical contexts, such as solving quadratic equations and working with irrational numbers.

Division of Infinite Series of Surds

$$\sqrt[n]{a \div \sqrt[n]{a \div \sqrt[n]{a \div \sqrt[n]{a \div} \ldots \ldots \ldots \infty}}} = \sqrt[(n+1)]{a}$$

Prove the identity:

Let us assume the division of entire series is equal to x

$$\sqrt[n]{a \div \sqrt[n]{a \div \sqrt[n]{a \div \sqrt[n]{a \div} \ldots \ldots \ldots \infty}}} = x$$

We can replace left hand side with x except first term for infinite series, the above series reduced as-

$$\sqrt[n]{a \div x} = x$$

Now, raising both side to the power of $\frac{1}{n}$

$$(a \div x) = x^{1/n}$$

$$\frac{a}{x} = x^{1/n}$$

And

$$a = x^{1/(n+1)}$$

$$x = \sqrt[(n+1)]{a}$$

If $a^{(n+1)}$, then, $x = a$

Example-1: if $a = k^3$ then find the value of the infinite division series

$$\sqrt{k^3 \div \sqrt{k^3 \div \sqrt{k^3 \div \sqrt{k^3 \div \sqrt{k^3 \div \ldots \ldots \ldots \infty}}}}}$$

Select the correct answer -

   i) k^3
   ii) k^2
   iii) k
   iv) None of the above

Solution:

As we know

$$\sqrt[n]{a \div \sqrt[n]{a \div \sqrt[n]{a \div \sqrt[n]{a \div \ldots \ldots \ldots \infty}}}} = \sqrt[(n+1)]{a}$$

Then we can write

$$\sqrt{k^3 \div \sqrt{k^3 \div \sqrt{k^3 \div \sqrt{k^3 \div \ldots\ldots\ldots\infty}}}} = \sqrt[2+1]{k^3}$$

$$= \sqrt[3]{k \times k \times k} = k \text{ Ans.}$$

Example-2: Find the value of the infinite division series

$$\sqrt{125 \div \sqrt{125 \div \sqrt{125 \div \ldots\ldots\ldots\infty}}}$$

Select the correct answer -

   i)   125

   ii) 25

   iii) 5

   iv) None of the above

Solution

$$\sqrt{125 \div \sqrt{125 \div \sqrt{125 \div \ldots\ldots\ldots\infty}}} = \sqrt[3]{125}$$

$$= \sqrt[3]{5 \times 5 \times 5} = 5 \text{ Ans.}$$

Example-3: Find the value of the given infinite division series

$$\sqrt{216 \div \sqrt{216 \div \sqrt{216 \div \sqrt{216 \div \ldots\ldots\ldots\ldots\infty}}}}$$

Select the correct answer -

   i)   6
   ii)   36
   iii)   216
   iv)   None of the above

Solution

$$\sqrt{216 \div \sqrt{216 \div \sqrt{216 \div \sqrt{216 \div \ldots\ldots\ldots\ldots\infty}}}}$$

$$= \sqrt[3]{216} = \sqrt[3]{6 \times 6 \times 6} = \textbf{6 Ans.}$$

Example-4: Find the value of the series

$$\sqrt{512 \div \sqrt{512 \div \sqrt{512 \div \sqrt{512 \div \ldots\ldots\ldots\ldots\infty}}}}$$

Select the correct answer -

    i)   8

    ii)   16

    v)   64

    vi)   512

Solution

$$\sqrt{512 \div \sqrt{512 \div \sqrt{512 \div \sqrt{512 \div \ldots \ldots \ldots \infty}}}} = \sqrt[3]{512}$$

$$\sqrt[3]{512} = \sqrt[3]{8 \times 8 \times 8} = \textbf{8 Ans.}$$

Example-5: Find the value of the series

$$\sqrt{64 \div \sqrt{64 \div \sqrt{64 \div \sqrt{64 \div \ldots \ldots \ldots \infty}}}}$$

Select the correct answer -

    i)   64

    ii)   16

    iii) 4

    iv) None of the above

Solution

$$\sqrt{64 \div \sqrt{64 \div \sqrt{64 \div \sqrt{64 \div \ldots \ldots \ldots} } } } \, \infty = \sqrt[3]{64}$$

$$\sqrt[3]{64} = \sqrt[3]{4 \times 4 \times 4} = \textbf{4 Ans.}$$

Example-6: Find the value of square root sum of infinite series

$$\sqrt{56 + \sqrt{56 + \sqrt{56 + \sqrt{56 + \cdots \ldots \ldots} } } } \, \infty$$

Select the correct answer -

   i)   8
   ii) 16
   iii) 64
   iv) None of the above

Solution

Let us assume the square root sum of infinite series is equal to x

$$\sqrt{56 + \sqrt{56 + \sqrt{56 + \sqrt{56 + \cdots \ldots \ldots \infty}}}} = x$$

We can replace left hand side with x except first term of the infinite series, the above series reduced as-

$$\sqrt{56 + x} = x$$

Now, squaring both the side of above equation, we have

$(56 + x) = x^2$

$x^2 - x - 56 = 0$

$(x - 8)(x + 7) = 0$

Here we can take positive value of x

$x = 8$

Thus, we got the value of infinite series, $x = 8$

Example-7: Find the value of x in given infinite series

$$\sqrt{x + \sqrt{x + \sqrt{x + \sqrt{x + \cdots \ldots \ldots \infty}}}} = 5$$

    i) 5
    ii) 10
    iii) 20
    iv) None of the above

Solution

We can write the above series as

$$\sqrt{x + 5} = 5$$

Now, squaring both sides
$$x + 5 = 25$$
$$x = (25 - 5) = 20$$

Thus, the value of $x = \mathbf{20}$

Example-8: Find the value of infinite series

$$\sqrt{7\sqrt{7\sqrt{7\sqrt{7}\ldots\ldots\ldots\infty}}}$$

Select the correct answer -

- i) 1
- ii) 7
- iii) 17
- iv) None of the above

Solution

Let us assume that, the square root of infinite series is equal to x

$$\sqrt{7\sqrt{7\sqrt{7\sqrt{7}\ldots\ldots\ldots\infty}}} = x$$

Then, we can reduced the above series as-

$$\sqrt{7x} = x$$

Now, squaring both sides

$$(7x) = x^2$$
$$x^2 - 7x = 0$$
$$x(x - 7) = 0$$

$$x = 7 \text{ or } x = 0$$

Thus, we got the value of infinite series, $x = 7$

Example-9: Find the value of following and select correct answer

$$\sqrt{19\sqrt{19\sqrt{19\sqrt{19}}}}$$

Select the correct answer -

i) $\quad 19^{1/4}$
ii) $\quad 19^{1/8}$
iii) $\quad 19^{1/16}$
iv) $\quad 19^{1/32}$

Solution:

Since, $\sqrt{a} = (a)^{(1/2)}$.

So, we have

$$\sqrt{19\sqrt{19\sqrt{19\sqrt{19}}}} = (19)^{(1/2).\,(1/2)\,(1/2)\,(1/2)} = (19)^{(1/16)}$$

Simplification Tricks For Exponents (Power of the Base).

Exponents, also known as indices or powers, indicate how many times a number (the base) is multiplied by itself.

Basic rules of exponent

Multiplication

$$a^m \times a^n = a^{m+n}$$

Division

$$\frac{a^m}{a^n} = a^{m-n}$$

Power of power

$$(a^m)^n = a^{mn}$$

Zero Exponent

$$a^0 = 1 \text{ (where } a \neq 0)$$

Negative Exponent

$$a^{-n} = \frac{1}{a^n} \text{ (where } a \neq 0)$$

Example-1 Which of the following is correct

   i) $(2^2)^3 = (2^3)^2$

   ii) $(2^2)^3 > (2^3)^2$

   iii) $(2^2)^3 < (2^3)^2$

Solution

$$(2^2)^3 = 2^{2\times3} = 2^6$$

And

$$(2^3)^2 = 2^{3\times2} = 2^6$$

Thus, the correct answer is (i) $(2^2)^3 = (2^3)^2$

Example-2 Find the value of x from equation $(2^2)^8 = (16)^{2x}$

   i) $x = 2$

   ii) $x = 4$

   iii) $x = 8$

   iv) $x = 16$

Solution

L.H.S. $(2^2)^8 = 2^{2\times8} = 2^{16}$

R.H.S. $(16)^{2x} = (2^4)^{2x} = 2^{8x} = 2^{8x}$

Now, comparing both the sides, we have

$$2^{16} = 2^{8x}$$

Here we can see that base are same, therefore the power will also be equal

$$8x = 16$$

Thus, $x = 2$ **Ans.**

Example-3 Find the value of x from given equation

$$9^x + 9^x + 9^x = (3)^{x+5}$$

     i) $x = 2$

     ii) $x = 4$

     iii) $x = 8$

     iv) $x = 16$

Solution

L.H.S. $9^x + 9^x + 9^x = 9^x(1 + 1 + 1) = 9^x(3)^1$

$$= 9^x(3)^1$$

$$= (3^2)^x.(3)^1$$

$$= 3^{2x}.3^1$$

$$= (3)^{2x+1}$$

R.H.S. $(3)^{x+5}$

Now, comparing both the sides, we have

$$(3)^{2x+1} = (3)^{x+5}$$

Here we can see that, the base are same, therefore the power will also be equal, thus we can reduce above equation as

$$2x+1 = x+5$$

Thus, $x = 4$ **Ans.**

Example-4: Find the value of $x = ?$

$$\sqrt{\sqrt{\sqrt{\sqrt{\sqrt{5}}}}} = 5^{(1/x)}$$

Select the correct answer -

i) $x = 5$

ii) $x = 4$

iii) $x = 16$

iv) $x = 32$

Solution

L.H.S. of the equation can be written as

$$\{\{\{\{\{5^{1/2}\}^{1/2}\}^{1/2}\}^{1/2}\}^{1/2} = 5^{(1/x)}$$

Applicable basic rule of power

$$(a^m)^n = a^{mn}$$

Therefore, above equation can be written as

$$5^{(1/2).(1/2).(1/2).(1/2).(1/2)} = 5^{(1/x)}$$

$$5^{(1/32)} = 5^{(1/x)}$$

Since, base are same on both side, therefore power will also be same

$$(1/x) = (1/32)$$

$$x = 32 \textbf{ Ans.}$$

Chapter – 14

Sutras in Vedic Mathematics with Applications

Vedic Mathematics is based on 16 sutras and 13 sub-sutras. They are the fundamentals behind the speed calculation in vedic maths. These Vedic sutras, are concise and easily memorisable aphorisms, provide systematic techniques for performing arithmetic operations, algebraic calculations, and geometric constructions with remarkable speed and simplicity.

These sutras are not only enhance computational accuracy but also promote the intuitive understanding and mental swiftness, making mathematics easy and enjoyable.

The 16 main sutras (aphorisms) and 13 sub-sutras (corollaries) along with applications in vedic mathematics.

S.No.	Sutra (Aphorisms)	Applications
1	**Ekadhikena Purvena** (*By One more than the Previous one*)	Square of numbers ending in 5
2	**Nikhilam Navatash Charamam Dashatah** (*All from 9 and the last from 10*)	Subtraction of Numbers, Finding complement of number from the base. Base multiplication, multiplication by 9, and division by the number 9.

3	**Urdhva-Tiryagbhyam** *(Vertically and crosswise -Criss-Cross)*	Multiplication of any number and multiplication of algebraic equations
4	**Paraavartya Yojayet** *(Transpose and apply)*	Division of number near power of 10
5	**Shunyam Saamya-Samuchhaye** *(When the "Saamya-Samuchhaye" is the same it is equated to zero)*	Solve algebraic equations, where the sum of terms is zero
6	**Anurupye Shunyamanyat** *(If one is in ratio, the other is zero)*	Simplifies solving equations where terms are in ratio, leading one term to be zero.
7	**Sankalana-vyavakalanabhyam** *(By Addition and by Subtraction)*	Solves simultaneous linear equations.
8	**Purana-Puranabhyam** *(By the completion or non-completion)*	Used for finding products and completing numbers.
9	**Chalana-Kalanabhyam** *(Differences and similarities)*	Deals with calculus, specifically finding derivatives and integrals using differences and similarities.

10	**Yaavadunam** *(Whatever the extent of its deficiency)*	Simplifies squaring numbers less than a power of 10
11	**Vyashti-Samanstih** *(Part and whole)*	Simplifies multiplication and division by decomposing into parts and whole.
12	**Shesanyankena Charamena** *(The remainders by the last digit)*	Used for finding remainders using the last digits.
13	Sopaantyadvaya-Mantyam *(The ultimate and twice the penultimate)*	Multiplication by 12 and deals with solving quadratic equations by specific manipulations of terms.
14	**Ekanyunena Purvena** *(By One less than the One before)*	Multiplication by 9
15	**Gunita-Samuchyah** *(The product of the sum is equal to the sum of the product)*	Simplifies polynomial factorization and equation solving.
16	**Gunaka-Samuchyah** *(The factors of the sum is equal to the sum of the factors)*	Another method for simplifying polynomial factorization and solving algebraic equations.

Sub-Sutras (Corollaries) & Applications

S.No.	Sutra	Applications
1	**Anurupyena** *(Proportionately)*	Used for finding proportional relationships. e.g. Working base multiplications.
2	**Shishyate Sheshangyah** *(The remainder remains constant)*	Used in division to find remainders consistently.
3	**Adhyamadhyenantya-Mantyena** *(The first by the first and the last by the last)*	Simplifies multiplication of binomials and polynomials.
4	**Kevalaih Saptakam Gunyat** *(By 7 and multiples of 7)*	Simplifies operations involving multiples of 7.
5	**Vestanam** *(By Osculation)*	Used for finding roots of equations through successive approximations.
6	**Yavadunam Tavadunikritya Varga cha Yojayet** *(Lessen by the deficiency and set up the square of the deficiency)*	Used for squaring numbers close to the base by reducing them to a base and correcting.
7	**Antyayoreva** *(Only the last terms)*	Simplifies multiplication and division (If last terms add up to 10).

8	**Antyayor Dashake'pi** *(Last terms of the sum)*	Focuses on the last terms of the sum for simplification.
9	**Samuchchayagunitah** *(Sum of the product)*	Used for simplifying sums and products in algebra.
10	**Lopanasthapanabhyam** *(By alternate elimination and retention)*	Simplifies complex equations by alternately eliminating and retaining terms.
11	**Vilokanam** *(By observation)*	Solving problems through keen observation and pattern recognition.
12	**Gunita-Samuchchayah Samuchchayagunitah** *(The product of the sum of the coefficients in the factors is the sum of the coefficients in the product)*	Used for polynomial expansions and simplifications.
13	**Dhvajanka** *(Flag digit)*	Involves specific flag digits for simplifying mathematical operations. e.g. Straight Division

These sutras and sub-sutras provide a comprehensive framework for a wide range of mathematical operations, from basic arithmetic to advanced algebra and geometry.

Chapter – 15

Swift Mental Calculations Tricks

Mental mathematics involves solving mathematical problems using mental calculations, without calculators, pens, or paper. It give emphasizes to the ability to perform arithmetic operations and solve mathematical problems quickly and accurately in your mind. This skill is often developed through practice. It will helps children to solve problems quickly using mental calculations, and enhance the quick decision-making skills in children.

In this chapter, we will first learn the techniques and then practice the simple mental mathematics for quick answers.

Mental Tricks of Addition

Example-1: What is 47 + 58?

Mental Steps: Break the numbers into tens and ones:

- 47 is 40 + 7, and 58 is 50 + 8.
- Add the tens: 40 + 50 = 90.
- Add the ones: 7 + 8 = 15.
- Combine the results: 90 + 15 = 105.

Practice the Following Additions

1) 37 + 46   2) 58 + 29
3) 63 + 18   4) 77 + 45
5) 82 + 39   6) 54 + 27
7) 91 + 14   8) 68 + 53
9) 76 + 22   10) 85 + 47

Mental Tricks of Subtraction

Example 2: What is 92 - 36?

Mental Steps: Break the numbers into tens and ones:

- Subtract the tens: 90 - 30 = 60.
- Subtract the ones: 2 - 6 = -4.
- Adjust the result: 60 - 4 = 56.

Practice the Following Subtractions

1) 92 – 47   2) 105 – 68
3) 84 – 29   4) 123 – 77
5) 150 – 89   6) 71 – 34
7) 98 – 46   8) 155 – 98
9) 60 – 23   10) 142 – 67

Mental Tricks of Multiplication

Example 3: What is 14×15?

Mental Steps: Break 15 into tens and ones

- $14 \times 15 = 14 \times (10 + 5)$
- Multiply 14 by $10 = 14 \times 10 = 140$, and
- Multiply $14 \times 5 = 70$.
- Add the results: $140 + 70 = 210$.

Practice the Following Multiplications

1) 7×8 2) 12×15
3) 6×9 4) 14×11
5) 5×16 6) 23×4
7) 8×13 8) 15×7
9) 9×12 10) 17×6

Mental Tricks of Division

Example 4: What is $65 \div 5$?

Mental Steps: Break 65 into tens and ones i.e. $(60 + 5)$

- Divide $60 \div 5 = 12$
- Divide $5 \div 5 = 1$
- Add the results: $12 + 1 = 13$.

Practice the Following Divisions

1) $84 \div 7$ 2) $144 \div 12$
3) $63 \div 9$ 4) $100 \div 5$
5) $56 \div 8$ 6) $72 \div 6$
7) $45 \div 9$ 8) $88 \div 11$
9) $120 \div 15$ 10) $98 \div 14$

Mental Tricks of Percentages

Example 5: What is 12.5% of 80?

Mental Steps: Since, we know the following facts:

100% of 100 is = 100/1 = 100
50% of 100 is = 100/2 = 50
25% of 100 is = 100/4 = 25
12.5% will be = 100/8 = 12.5.

- Now, to find out 12.5% of 80
- Divide by 8
- $80 \div 8 = 10$.

Practice the Following Percentage

1) 20% of 150   2) 35% of 80
3) 50% of 60   4) 75% of 120
5) 10% of 250   6) 40% of 90
7) 25% of 200   8) 60% of 50
9) 85% of 160   10) 15% of 40

These steps help simplify mental calculations and improve speed.

Chapter – 16

Exercise For Practice

1. Practice the following using vedic tricks

S.No.	I. Addition		II. Subtraction
1)	3,578 + 6,432	1)	7,654 - 3,218
2)	4,567 + 5,678	2)	9,876 - 4,321
3)	2,345 + 4,567	3)	5,432 - 2,678
4)	1,234 + 3,456	4)	8,910 - 1,234
5)	2,345 + 6,789	5)	6,789 - 3,456
6)	4,321 + 5,432	6)	4,321 - 1,876
7)	3,210 + 4,321	7)	9,012 - 4,567
8)	1,456 + 2,345	8)	7,890 - 2,345
9)	7,890 + 8,901	9)	5,678 - 2,345
10)	2,678 + 3,789	10)	8,234 - 3,789
11)	5,432 + 6,543	11)	6,543 - 2,456
12)	3,456 + 5,678	12)	4,567 - 1,234
13)	4,567 + 6,789	13)	9,345 - 5,678
14)	1,234 + 2,345	14)	7,123 - 3,789
15)	6,789 + 7,890	15)	5,890 - 2,456

2. Solve the following using vedic tricks

S.No.	I. Multiplications		II. Divisions
1)	$3,578 \times 6,432$	**1)**	$4,568 \div 12$
2)	$4,567 \times 5,678$	**2)**	$7,890 \div 45$
3)	$2,345 \times 8,910$	**3)**	$6,543 \div 23$
4)	$1,234 \times 9,012$	**4)**	$8,910 \div 37$
5)	$6,789 \times 7,123$	**5)**	$5,432 \div 19$
6)	$4,321 \times 7,654$	**6)**	$9,876 \div 54$
7)	$5,432 \times 6,543$	**7)**	$3,456 \div 28$
8)	$2,678 \times 9,876$	**8)**	$7,123 \div 31$
9)	$8,901 \times 7,890$	**9)**	$2,345 \div 67$
10)	$3,456 \times 5,678$	**10)**	$8,234 \div 42$
11)	$6,543 \times 8,765$	**11)**	$4,321 \div 56$
12)	$7,890 \times 9,012$	**12)**	$5,678 \div 29$
13)	$5,678 \times 4,321$	**13)**	$9,012 \div 38$
14)	$2,345 \times 7,890$	**14)**	$6,789 \div 75$
15)	$9,876 \times 1,234$	**15)**	$3,890 \div 22$

3. Solve the following using vedic tricks

I. Square Roots

1. $\sqrt{1,600}$
2. $\sqrt{2,401}$
3. $\sqrt{3,600}$
4. $\sqrt{4,841}$
5. $\sqrt{6,400}$
6. $\sqrt{8,100}$
7. $\sqrt{9,604}$
8. $\sqrt{2,500}$
9. $\sqrt{3,024}$
10. $\sqrt{4,084}$

II. Cube Roots

1. $\sqrt[3]{1,000}$
2. $\sqrt[3]{1,331}$
3. $\sqrt[3]{2,000}$
4. $\sqrt[3]{3,375}$
5. $\sqrt[3]{4,096}$
6. $\sqrt[3]{5,832}$
7. $\sqrt[3]{7,000}$
8. $\sqrt[3]{8,000}$
9. $\sqrt[3]{9,261}$
10. $\sqrt[3]{10,648}$

4. Solve the following multiplications swiftly using vedic tricks

1)	11×35	**2)**	67×47
3)	72×78	**4)**	95×95
5)	35×45	**6)**	55×55
7)	864×999	**8)**	27×101
9)	11111×11111	**10)**	111×423

5. What is difference between 9×9 matrix and vedic square?

6. Create magic square of 3×3 and it's complement square?

7. How many Sutra (aphorisms) and Sub-Sutra (corollaries) are there in Vedic Mathematics?

8. Explain the Vedic Sutra "All from 9 and the Last from 10", with its applications.

9. Explain base and transpose and applications in number division?

10. Write down 5-best tricks you have lean and share with your friends and family.

Author

Author Connect

Hello, readers I, invite you to my educational website *www.sparktriangle.com*, where you will find diverse collection of my published works and other valuable resources dedicated to supporting students and professionals in achieving their career aspirations.

I have seamlessly blend my experience from diverse fields with a passion for writing, crafting engaging narratives with readers globally. In 2023, I published my first book, "*Miracles of Vedic Tricks for Mental Maths: Unleashing the Power of Lightning-Fast Calculation*" and I am committed to continuing this journey of writing and sharing with readers worldwide. I am also writing on juvenile and non-fictions genres.

I am working hard on my upcoming projects under juvenile and non-fiction categories, they are deeply inspired by the *Shrimad Bhagavad Gita*. I believe it will be a game-changer for children and young professionals. *I wish you all the success in your future endeavours.*

Explore More from *MAHAVIR*

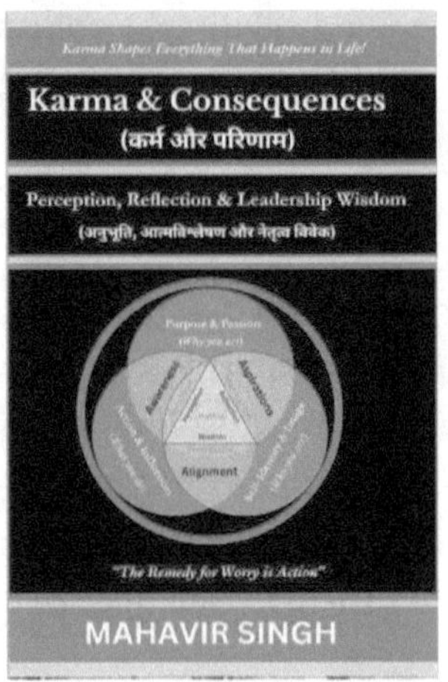

Online Available as (*Print Book, eBook* and *Audio Book*)

Feedback and suggestions for improving this book are greatly appreciated.

Visit to Learn More

https://sparktriangle.com/

Feel free to Write & Connect Us

"Imagination is more important than knowledge. For knowledge is limited, whereas imagination embraces the entire world, stimulating progress, giving birth to evolution."

- Albert Einstein